Design for Flying

Design for Flying

2nd Edition

David B. Thurston

TAB Books
Division of McGraw-Hill, Inc.
New York   San Francisco   Washington, D.C.   Auckland   Bogotá
Caracas   Lisbon   London   Madrid   Mexico City   Milan
Montreal   New Delhi   San Juan   Singapore
Sydney   Tokyo   Toronto

pbk   1 2 3 4 5 6 7 8 9 10   DOH/DOH 9 9 8 7 6 5 4
hc   1 2 3 4 5 6 7 8 9 10   DOH/DOH 9 9 8 7 6 5 4

Library of Congress Cataloging-in-Publication Data

Thurston, David B.
    Design for flying / by David B. Thurston.—2nd ed.
        p.   cm.
    Includes index.
    ISBN 0-07-064557-4   ISBN 0-07-064559-0 (pbk.)
    1. Airplanes—Design and construction.   I. Title.
TL671.2.T57 1994
629.134'1—dc20   94-10828
                                        CIP

Acquisitions editor: Jeff Worsinger
Editorial team: Joanne M. Slike, Executive Editor
                Lori Flaherty, Managing Editor
                Pam Reichert, Book Editor
Production team: Katherine G. Brown, Director
                Lisa M. Mellott, Coding
                Jan Fisher, Desktop Operator
                Joann Woy, Indexer
Design team: Jaclyn J. Boone, Designer   GEN3
                Brian Allison, Associate Designer   0645590

Contents

Preface

Design for Flying was originally written in response to the many inquiries and questions I received over the years concerning aircraft handling characteristics, operational procedures, and detailed design. Since the answers are of great interest to a broad cross section of the flying public, this book was prepared with a minimum of mathematics—requiring nothing beyond simple algebra to understand the most complex of the few formulas presented.

Important design criteria are reviewed by running through a flight profile from takeoff to landing, with the critical requirements for each phase of flight presented at three levels: (1) why aircraft must fly as they do to satisfy pilot and FAA handling requirements; (2) what design features make the airplane behave as required; and (3) what design criteria are necessary to establish these characteristics. The first two levels are of interest to everyone concerned with aircraft, while the technical explanation may be skipped if it is outside any reader's interest. By exploring the relationship between aircraft design features and flight characteristics, *Design for Flying* invites each of you to examine the effects of your own piloting techniques on aircraft response.

As a matter of interest and to provide additional background information, the history and development of essential criteria are briefly reviewed along with the various phases of design. Although few basically new concepts have been introduced into the aircraft field since 1911, the old ideas are modernized from time to time and, where relevant, their effect on current aircraft is noted. For those of you interested in digging further into technical items or designing your own airplane, useful design and maintenance references follow each chapter.

Further, if nontechnical readers wish to change the speed or climb range of their propeller, equip their airplane for IFR flight, position avionics antennas, or modify existing equipment items, *Design for Flying* tells what must be done and how to go about doing it—information that should save every aircraft owner both time and money.

Design for Flying is a general aviation encyclopedia, providing an excellent background and reference source for ground and flight school instruction, technical schools and individuals interested in practical design, and all pilots curious about how aircraft is designed and what makes planes fly as they do.

In revising the complete text for this second edition, I was amazed at how much and in how many ways general aviation has changed over the past 17 years since *Design for Flying* was first written. During the 1970s, the production of general aviation aircraft was about 10 times the present volume, with airplanes costing about one-quarter what they do today. Adding the present cost of fuel, insurance, and maintenance work, it is no wonder general aviation growth has slowed to a minimum level.

But the basic physics of flight and related aerodynamics remain constant regardless of sales volume. Although the many references for each chapter might seem dated, upon review they remain the latest and best available thanks to the current limited status of personal aircraft production and NASA's present interest in outer space rather than subsonic aerodynamic development. We still have not produced that 5-to-1 speed ratio airplane capable of cruising at 200 mph and landing at 40. But it cannot be far away!

Naturally, during 53 plus years of active design experience, many people have influenced my interest in the aircraft field. In addition to expressing my sincere appreciation to everyone, I would particularly like to thank the following collaborators who were of assistance in getting this book together:

The publishers and editors of *Aero Magazine*, first Marvin Patchen and then Will Hunt, who over the years encouraged and supported my interest in writing and publishing semitechnical aviation articles, some of which have been re-edited with their approval to form the basis for parts of this book. Two capable airmen, Richard A. Bradley and Walter H. Prentice, both of whom are competent Certified Ground and Flight Instructors holding ratings for commercial and instrument training. In addition, Dick is a corporate jet captain and Walt is a qualified aerobatic instructor. Their comments and guidance have made *Design for Flying* a more useful and practical book. My friend for many years, a former test pilot and fellow engineer, Hank (Franklin T.) Kurt, who encouraged preparation of this book and offered many helpful suggestions prior to his terminal illness. The many aircraft sales and public relations people who responded rapidly and enthusiastically to requests for photographs and other product information. Brian Alward who carefully placed the revised text on computer. And last, but by no means least, my wife, Evelyn, who had the perseverance to translate my handwriting into the original typed manuscript.

David B. Thurston
Design for Flying
January 1994

1

Basic design considerations

Early aircraft

Ever since humans learned to stand erect they have been preoccupied with a desire to leave the ground and fly in a controlled direction. If we assume that the surviving legends, cave decorations, and tapestry details have some basis in fact, humanity's early urge to fly was a small but constant contributor to population reduction.

Because flight is the visible result of many phases of applied physics, positive advancement must be keyed to state-of-the-art engineering, mechanical design, and fabrication techniques. When we review the many early and unsuccessful attempts at flight, it becomes obvious that controlled flight with wings depended on some theory of lift of inclined planes (wings) and so had to await the required mathematical equations. As a result, balloons—air-displacing vehicles—logically provided the first real demonstration of atmospheric flight. As is well known, the Montgolfier brothers, French manufacturers of fine paper, are generally credited with the first public balloon flight on September 19, 1783 using hot air for lift. However, three weeks earlier the distinguished French physicist Jacques Charles (discoverer of Charles's law) had secretly launched an unmanned hydrogen-filled balloon that flew for some distance to the terror of all local peasantry.

Typical of aeronautical development, a field in which any advance is eagerly pursued by all participants, balloons rapidly proliferated. I am always amazed at the amount of time, money, and enthusiasm people seem to find today for new development projects of every sort, and apparently this was just as true in the late eighteenth century. Balloon flights soon crossed the English Channel and the Alps, thrilled crowds in North and South America, and naturally, were applied at once to military observation, where balloons were successfully employed as recently as World War II.

Although many fortunes were spent attempting winged flight, nothing new developed until a text covering the physics of fluid motion was published by Horace Lamb in 1879. This work, combined with existing theories, such as Bernoulli's principle, finally permitted rational calculation of the lift and center of pressure of flat and cambered surfaces. In 1885, professor John J. Montgomery of Santa Clara, California successfully flew a manned glider with aileron control, as shown in Fig. 1-1, subsequently publishing his equations for lift of flat and curved planes in 1890. Many other mathematicians were at work during this same period, with the result that we were scientifically more or less ready for powered flight by 1900 but lacked an adequate engine.

Ailerons shown displaced to
indicate amount of travel

Aileron motion obtained by
pilot rotation of saddlebar

1-1 *Montgomery's aileron-equipped glider.*

Every quantum jump in progress requires some fundamental technical advancement. Just as manned space flight awaited development of the solid-state computer, manned atmospheric flight was dependent on an efficient powerplant of some sort. This contribution was made by the Wright brothers through development of a gasoline engine and drive system superior to any then available.

Certainly another fundamental achievement made by the Wrights was their technical analysis of aerodynamic design before starting full-scale construction and flight demonstration. Most of the early pioneers were trial-and-error specialists and the law of survival of the fittest surely applied. As a result, any single flight completed without injury or fatality contributed to aeronautical knowledge. Fortunately, because of the many participants, aircraft and engine design and performance developed despite these ordeals.

Lateral control is basic to successful flight. It is a crucial control for crosswind operation near the ground and is essential for all coordinated flight maneuvers. With the exception of early designs patented by Montgomery of the United States and Louis Pierre Mouillard of France, early experimenters were not aware of the need for lateral control. In fact, most designers were so involved in becoming airborne that any requirement for a controlled turn was secondary to getting off the ground for a measurable period of time. And yet lateral control was to become a problem as basic as that of structural design and efficient power. Every current competition for man-powered flight requires completion of a figure-eight flight path, providing recognition of the basic difficulty of controlled turning flight.

The fact that the Wrights and Glenn Curtiss did realize the fundamental need for lateral control partially explains their early success. Using a different lateral control approach from the Wright's wing warping system, Curtiss developed separate ailerons initially hinged about the wing's leading edge. These ailerons, located at the tip of each wing, were successfully demonstrated on the 1908 June Bug and were subsequently used on Curtiss's designs in the arrangement shown in Fig. 1-2.

Similarly, until the Wright brothers installed rudder control in late 1904 little thought had been given to use of vertical surface for turn coordination and assistance. As a matter of interest, the first recorded circular flight under power was made by Wilbur Wright on September 20, 1904, who used wing warping plus rudder control. But the added drag of turning flight combined with marginal thrust and poor pilot technique resulted in many crashes for many pilots before both power and skill were developed to a level permitting practical and safe controlled flight.

The art of powered flight developed very rapidly and by 1911 aircraft of all configurations had been built and flown. In fact, with the exception of fully retractable landing gear, no new arrangement of wings, tail surfaces, control system, alighting gear (land or water), or reciprocating powerplant has been developed since. By 1911, Curtiss had flown tricycle landing gear, developed a wheel control with integral throttle, produced relatively high-powered/low-weight engines, and designed amphibious floats with externally retractable wheels. In addition to correctly analyzing the criteria for a coordinated turn and developing the required aerodynamic configuration, the Wright brothers had originated the skid landing gear and the catapult launching system (similar to

1-2 *Curtis 1910 biplane showing the mid-gap ailerons and shoulder yoke control.*

present aircraft carrier practice), improved engine reliability and propeller efficiency, and refined their structural design procedures sufficiently to satisfy basic military procurement requirements.

By 1911, other powered aircraft experimenters had developed tandem and four-wheeled landing gear as well as the tail-wheel-type. A tractor monoplane had flown across the English Channel, destroying forever the isolation of the British Isles. Paul Cornu had designed and built a model V/STOL (vertical/short takeoff and landing) helicopter having a wing, rotors that could be deflected to permit forward flight, and a skid landing gear. Canard, tandem, monoplane, biplane, and triplane wing arrangements had been flown with varying degrees of success. Wheel, stick, shoulder yoke, and rudder bar were used in arbitrary combinations for turning and climbing flight. Propellers were located at the fuselage nose and tail, as well as on aircraft wings. And rotary and reciprocating air-cooled and water-cooled gasoline engines had been flown, as had wooden and metal-bladed fixed and variable pitch propellers.

According to the 1910 text *Aerial Navigation of To-Day*, Thomas Edison had recently predicted the helicopter would be the flying machine of the future, designers considered the monoplane superior to the biplane for high-speed flight, Russia had stated her right to fire at any balloon crossing her frontier, and European officials were concerned about possible "smuggling by aeronauts." So not much has happened in the past nine decades of aeronautical development that was not tested or anticipated by 1911.

The evolution of design specifications

The improvement of aircraft design from 1911 concepts to current levels has paralleled the combination of practical experience gained from flight plus the refinement of engineering techniques realized from structural test, flight test, and materials development programs.

Despite the flight evaluation of a multitude of design configurations by 1911, many of these aircraft would have been extremely dangerous to fly if adequate power had been available. Airframe structural design was also an unknown art conducted largely by trial and error, with frequently fatal results due to in-flight structural failures. Further, nothing indicates that early designers considered wind gusts or their effect on flight loads and flight-path control during the approach. This gap in essential background analysis is primarily because flight was a new art, which every experimenter and adventurer wished to savor immediately; rational solutions simply could not be realized within the time and funding limits available. In many ways, the aviation period from 1908 through 1912 was similar to the California Gold Rush of 1849; everyone wanted to get into the act and by any route possible.

Due to poor evaluation of flight requirements, many stall and skid crashes were never explained or fully understood during this early development period. However, through data obtained from a number of such accidents the Army Signal Corps gradually evolved an aircraft design procurement specification, with similar requirements established by the Navy for its air arm.

As would be expected, military procurement was the principal support of the struggling aircraft industry through World War I, with accompanying design specifications growing in length and detail as more was learned about aerodynamics and high strength-to-weight ratio structural design. When the fighting was over, so was the first peak of aviation growth. The marketplace was flooded with surplus military aircraft, notably the Curtiss Jenny, the DH-4, and similar aircraft that could be used by sports enthusiasts and barnstorming pilots, a sales pattern unfortunately repeated after World War II. When vacant fields of reasonable size started to sprout tandem-seated biplanes all across the United States, the Department of Commerce was appointed as licensing agent for aircraft and pilots, with authority to draft basic requirements for commercial aviation in the United States—and so established the foundation for our present Federal Aviation Administration (FAA).

From 1920 to 1929, many prize award flights were undertaken both here and abroad, some between great distances, as in Lindbergh's May 1927 transatlantic flight from Roosevelt Field to Paris. The airframe, engine, and instrument design requirements specified by the competing pilots and financing syndicates brought about major improvements in aircraft quality and reliability. Stability became of increasing importance during record flights, while the night mail runs indicated the great need for blind flight instrumentation and icing protection in addition to inherent stability.

Whenever flight evaluation demonstrated that a design modification, new device, or flight performance requirement directly contributed to improved flight safety, it would eventually be incorporated into the Airworthiness Requirements. These specifications were controlled by the Bureau of Air Commerce through the 1930s, then the Civil Aeronautics Administration, the Federal Aviation Authority, and now the Federal Aviation Administration acting as an arm of the Department of Transportation. And so we can see that aircraft design regulations developed from early Signal Corps days to the present through a rather obvious but natural Darwinian process of evolution.

Lightplane design responsibility

Unless they were in the unusual position of having a production order for a specific airplane design or they were employed by one of the "big three" private aircraft manufacturers, designers of new lightplanes frequently found no market for their products. In that case, prototype development time and money had been unfortunately wasted, and the sponsoring company could be forced out of business. This scenario has been repeated all too frequently in the lightplane business and explains why bankers take a basically dim view of financing new ventures in the private aircraft field. It also partially explains why manufacturers are hesitant to introduce radical designs, a subject reviewed in some detail in chapter 3.

The successful airplane company must offer the market an airplane of sound structural and mechanical design plus quality construction. The designer's and the manufacturer's reputations are probably more closely related to product integrity in the aircraft industry than in any other field. The explanation is quite fundamental: people's lives are at stake every time the product is used. A good-performing, easily handled, structurally sound airplane is rapidly accepted and establishes the reputation of the design and manufacturing company. Good news travels fast in the aircraft industry because a pilot in New England today can be in Tennessee tomorrow and Arizona the day after. Of course, bad news travels at the same speed and seems to grow with time and distance as well. So designers must be constantly aware of their responsibility to produce a sound, useful airplane. Let us see how these goals can be realized.

To improve operational safety the designer must eliminate as many features as possible that could lead to pilot error in flight or on the ground. To achieve this goal basically requires:

- An aerodynamic arrangement offering ample stall warning through buffeting and possibly "soft controls," which are apparent 5 to 10 mph above the stall speed (soft controls means a noticeable reduction in control forces).
- A stable airplane that can be readily trimmed during all phases of flight operation.
- A rate of climb exceeding 700 ft/min.
- A well-designed and tested structure.
- A proven, reliable powerplant.
- An automatic alternate source for engine intake air.
- Simple, "foolproof" engine-starting and fuel-delivery systems.
- Excellent pilot visibility during taxi and flight.
- A relatively quiet cabin to reduce pilot fatigue during long trips.

Features such as these, combined with modern radio navigation, represent virtually the entire advance in general aviation design since 1922. Today's personal aircraft do not fly any faster or more efficiently than yesterday's aircraft. The difference is that they can be comfortably navigated from point A to point B in relative safety, they have a reliable engine, and they are easy to handle.

Although flight safety is a paramount consideration, designing for low maintenance is also a prime requirement for a successful airplane. Nothing is more frustrating to an owner, or damaging to the industry, than an airplane that becomes a vast sinkhole for hard-earned (after-tax) dollars. The designer can be of great help here by:

- providing an efficient structure, one that is rugged without unduly penalizing useful load;
- eliminating complex hydraulic and electrical systems by using mechanical operation of flaps and landing gear;
- developing a clean design that obtains high cruise with fixed, faired landing gear;
- locating batteries, relays, brake system reservoirs, pitot lines, etc., in accessible areas;
- providing an instrument panel that is readily removable for service or inspection;
- using circuit breakers instead of fuses;
- using proven components and accessories that are backed by reputable companies.

These maintenance items might seem obvious; yet have you ever tried to find the connection to your vacuum system air filter, fill your battery, or find that hydraulic system slow leak that keeps staining the belly skins? Over a period of time, poor detail design can cost much more than the initial purchase price of any airplane. Ideally, many owners should be able to perform more of their own aircraft maintenance and clean-up work within permissible FAA service limits. For example, flight test data show that both speed and rate of climb can be increased by putting wheel fairings on fixed-tricycle landing gear. While some gain would be expected, the level of improvement was found to approach that of a fully retractable landing gear. A fixed landing gear means lower initial cost and maintenance plus one less opportunity for pilot error, since you can't land gear-up if the gear doesn't retract. These are the reasons most small aircraft have fixed gear equipped with wheel fairings.

The most popular aircraft

As an introduction to our study of why flight requirements actually design the successful airplane, consider just how a few basic design features relate to modern single-engine aircraft.

If success is measured by ownership popularity, the winner by a factor of at least 2 is the Cessna line of single-engine, fixed gear, four-place airplanes. Typical of this series is the Cessna 172 Skyhawk of Fig. 1-3.

More than 35,000 of these aircraft had been built when production was terminated during 1986. This design has ample stall warning and probably is the most stable of any airplane in its weight and speed class. In fact, I believe the solid feel of the single-engine Cessna models is a major reason for their popularity; they are stable flight platforms and so inspire pilot confidence.

Why is this so? To be stable in hands-off trimmed flight, which is referred to as *stick-free stability,* an airplane must have a fairly long tail arm (the distance from the center of gravity to the center of lift of the horizontal or vertical tail).

The horizontal tail should also be rather large to provide the rapid damping necessary to cancel pitch oscillations caused by turbulent air or pilot action. Cessnas have these features as shown in Fig. 1-3.

1-3 *Cessna 172 Skyhawk.*

Responsive controls are necessary at stall speed, during approach, and for operation under moderate to strong crosswind conditions. This is particularly true of the ailerons, so plenty of aileron area is required; while tapered outer wing panels should be used to reduce roll damping and thereby increase rate of roll for improved lateral control. Again, these features are on the Cessna models.

The high wing configuration provides excellent downward visibility in flight, although side vision tends to be restricted in a turn. The accepted technique to overcome this condition is to raise the wing on the intended turn side and look around before starting to bank. However, this potential blind spot is annoying in terminal areas and has undoubtedly been responsible for many "in-pattern" accidents. The larger 180-hp Cessna Cardinal has much improved pilot visibility in turns, plus much more elbow room and rate of climb, resulting in a generally superior airplane compared with the 172 Skyhawk—but has a superior price as well. At any rate, most owners must prefer downward visibility over turning visibility because Cessna single-engine sales continue to grow.

Tricycle landing gear is much easier to handle on the average runway than tail-wheel-type gear, especially in crosswind conditions. All major production light aircraft have this feature.

By way of satisfying our maintenance items, the Cessna 172 has fixed, faired landing gear and no hydraulic system other than the main wheel brakes. The fixed-pitch propeller simplifies flight operation because there is no pitch control requiring pilot attention, while the Lycoming 160-hp (172 Skyhawk) and 180-hp (Cardinal) engines are proven workhorses of record with long hours of operation between major overhauls. Due to the low position of the fuselage floor when parked, the Cessna 172 and Cardinal are both easy to enter and service.

The only real drawback found with the Cessna high-wing configuration occurs during preflight inspection of the fuel tanks, which are located high above ground in the wing panels. However, this is where the fuel should be and external steps can be obtained as an option to ease the fuel inspection problem.

Although the actual contribution of active sales and service departments in relation to total sales volume cannot be fully assessed, it is quite evident that the Cessna line of single-engine personal aircraft satisfy most of the design and maintenance criteria, and Cessna has the sales figures to prove it.

The number-two production position belongs to the Piper Cherokee line of low-wing, tricycle gear aircraft shown in Fig. 1-4. This design has a slightly shorter tail arm than the Cessna 172. It also has an all-movable horizontal tail that can be subject to annoying minor trim changes, apparently caused by local variations in airflow as well as by slight wear and play in the control system linkage. The combination of shorter tail arm and all-movable tail result in somewhat reduced, although adequate, longitudinal stability for the Cherokee series compared with the Cessna fixed stabilizer design. While the basic Cherokee rectan-

1-4 *1976 Piper Cherokee Cruiser.*

gular wing tends to reduce roll rate and crosswind response, the later Piper Warrior and Archer II designs have tapered outer panels that improve aileron effectiveness to meet or exceed Cessna levels.

The Cherokee low-wing arrangement provides two features everyone admires: visibility in the pattern and easier wing fuel tank inspection. The low wing also offers a convenient structure for mounting the main gear, while the Piper oleo struts are better shock absorbers than the Cessna spring steel main gear legs. The low wing and related support structure also combine to offer slightly superior crash protection during off-field landings, although a study of such accidents indicates the difference in fatalities favoring low-wing aircraft is much lower than might be expected.

Despite the many favorable aspects of a low-wing location, all such designs share a common problem—difficulty of cabin entry. Unless a sliding canopy is used, climbing through a side door positioned above a wing is difficult for anyone who is not a natural acrobat. And even the canopy approach might not be appreciated by women in skirts. However, the Cherokee cabin is roomier than that of the Cessna 172, making up for the more difficult entry.

The basic structure of the Cherokee models, particularly at the wing to fuselage connection, is slightly heavier than that of the Cessna singles. This results in less skin panel noise during turbulent conditions and, unfortunately, also in somewhat less useful load for the 150-hp Cherokee Cruiser compared with a Cessna 172 of similar power.

As a final comparison item, the Cherokees might tend to float a bit in ground effect and so are a bit more difficult to land than the high-wing Cessnas, particularly when approach speeds are a few miles per hour high.

Stepping back and weighing the various positive and negative features of each design, it is apparent that when given equivalent performance, powerplant reliability, maintenance requirements, and price, owners prefer an airplane offering ease of entry, slightly greater longitudinal stability, and downward visibility in comparison with one having superior visibility in the landing pattern but with somewhat less longitudinal stability, useful load, and ease of entry.

On balance there is little difference between the Cherokee and Cessna singles; both configurations provide excellent aircraft meeting most of the design and maintenance criteria outlined. But another manufacturer also has a very popular single-engine model, although of a very different breed. This airplane is the Beech Bonanza of Fig. 1-5. In production since 1947, this high-performance airplane is the acknowledged Rolls Royce of the general aviation market.

I remember attending the Cleveland Personal Aircraft Exhibition in 1946 with some Grumman associates and seeing the prototype Bonanza on display at a basic introductory price of $6900. I was then project engineer for a postwar personal aircraft design Roy Grumman believed would appeal to the many Navy pilots who had flown our rugged Grumman aircraft in the Pacific theater.

1-5 *Beech Bonanza V35B.*

After seeing the Bonanza and comparing it with our similar design priced at $8100, we decided Grumman's military overhead precluded competition in the private aircraft market, and so our new design was promptly dropped. It was a nice airplane that performed well with standard controls and with the briefly popular two-control system (the ailerons and rudder were interconnected to a control wheel but the elevator control action remained separate). As things developed the Bonanza never sold for $6900, and for many years I have wondered if Grumman should not have entered the lightplane market back in 1946. Beech unknowingly eliminated potential competition by posting an unrealistically low list price. The Bonanza remains in production today, but the equipped price can easily exceed 40 times the proposed 1946 list price.

Since first production in 1947, Beech has produced more than 20,000 Bonanza aircraft of various model designations. This design is popular with the more experienced pilot because it combines speed with a proven structure and excellent handling characteristics. The Bonanza is a superb instrument airplane in the hands of a qualified pilot who flies regularly and fills a definite and specialized segment of the single-engine private airplane market.

Although requiring more pilot attention and maintenance than either the leading Cessna and Piper aircraft, the Bonanza meets our basic design and most of the maintenance criteria. Owners pay a premium initially and for each flight hour in exchange for increased climb and cruising speed compared with the less sophisticated Cessna 172 and Cherokee aircraft. However, due to a combination of service reputation, declining dollar, and a well-organized owners association, the Bonanza enjoys an unusually high secondhand market value.

Many factors in addition to design contribute to customer acceptance. Among these are effective sales, spare parts, and service organizations; the location of qualified maintenance and overhaul facilities in each sales area; a well-finished interior; an active secondhand market; and the known fact that pilots tend to purchase the type of airplane in which they learned to fly. Continued engineering development is also very important. For example, the 180-hp Mooney Ranger has been developed into the 270-hp, higher-performance M20/TLS model of Fig. 1-6, offering a new level of performance to former Mooney customers and opening a new sales market for the company. Similar expanded product lines and sales markets are also common to Beech, Cessna, Piper, and Commander private aircraft.

Paul Bowen for Mooney Aircraft

1-6 *The 270-hp Mooney M20/TLS.*

According to the most recent AOPA (Aircraft Owners and Pilots Association) tabulations, the average private pilot only flies about 45 hr per year. Therefore, it's essential that aircraft designed for the private flying market be easily handled in all flight conditions. To summarize basic design and at the risk of seeming repetitive, criteria require:

• An airplane to be stable in flight at a given trim setting.
• Exhibit no peculiar or demanding stall or ground handling characteristics.

- Be easily controlled in the approach.
- Above all, exhibit positive longitudinal stability characteristics, as demonstrated by rapid damping of pitch oscillations in turbulent air and by positive stick forces (increasing control stick load with increasing pitch attitude).

Of course, the structure must be well tested and of proven design, while the powerplant should be relatively simple to operate, free of restricted (red band) rpm settings on the tachometer dial, and not subject to induction system or carburetor icing. In other words, the private airplane must be simple to fly.

References to other design material can be found in most chapters. These references are endnotes. Please realize that all examples presented are intended to illustrate a point or problem and, as such, might not necessarily represent precise engineering solutions—particularly since all calculations are simple algebra or lower level mathematics.

Within these guidelines, I'll cover aircraft design, starting with the origin of structural loads, then review the ever-continuing search for utility as represented by special aircraft configurations, and finally getting to the root of airplane design through a study of desirable operational requirements and how they are realized. The balance of the text illustrates why various design options must be combined or traded off to satisfy these needs, while also meeting FAA design and type certification criteria. Our aircraft are designed, developed, and refined through just such a process, usually a two- to three-year period.

2

The origin
of loads

Private aircraft are certified in the United States under FAA Federal Air Regulations (FAR) Part 23 Airworthiness Standards.[1] These design requirements are accepted by practically every country in the world, although some foreign agencies (particularly in Australia and France) reserve the right to review design and flight data for imported aircraft before issuing an airworthiness approval. As part of a reciprocal agreement arranged at the Department of State level, the FAA accepts aircraft design standards of many other governments upon an "equivalent basis" for import of private aircraft into this country.[2]

FAA specifications cover applied loads, airplane design, and flight test requirements in great detail. I first became acquainted with Air Commerce Manual 04, entitled *Airframe Airworthiness*, in 1939. At that time, ACM 04 cost 15 cents and totaled 140 pages that included the following:

- design terms and definitions;
- specific design requirements;
- static and flight test procedures;
- aerodynamic and applied loads calculations;
- weight and balance calculations; and
- aircraft modification and licensing procedures.

The 1993 regulations cover the same ground with six different parts costing about $65 and totaling more than 500 pages, not including any recommended procedures. In fact, Part 23 lightplane design requirements alone total more than 140 pages—so the quality and safety of small aircraft should have improved if the quantity of design specifications is any criterion.

Before proceeding further, a few simple definitions are helpful in explaining design load terminology. The *load factor* refers to the number of times the applied loading exceeds the 1-g weight. For example, a 10-lb radio subjected to a 4-g load factor applies a load of 40 lbs to its supporting structure. So the applied load is 40 lbs. The load factor can also be referred to as an *applied load factor*, a *limit load factor*, or an *ultimate load factor*. The applied and limit load factors represent the actual design loads experienced during operation divided by the weight at 1 g. Thus, we obtain a load factor of 4 for our radio support example.

The ultimate load factor is obtained by multiplying the limit load factor by a 1.50 margin of safety. So the 4-g limit load factor becomes a 6-g ultimate load factor, while the 40-lb limit load becomes a 60-lb ultimate load acting on the support structure. We also have *maneuvering* and *gust* load factors, which can be referred to as *limit maneuvering, limit gust, ultimate maneuvering, or ultimate gust* load factors. In each case, the 1.50 factor of safety margin changes limit to ultimate, and the ultimate loads are used to design the aircraft structure—creating a 50% overload safety margin.

Airframe is another definition of interest. The airframe is the basic aircraft structure designed to support the applied loads. These loads include the wings, tail surfaces, fuselage or hull, cowling, engine mount, control system, and landing gear. When all airframe components are assembled into one structure they support the powerplant, passengers, baggage, and fuel. So whenever the word airframe is used, it refers to the load-carrying structure of the airplane, thereby excluding such items as the engine, propeller, powerplant controls, instruments, radios, hydraulic and electrical systems, and any special equipment such as oxygen or pressurization systems. This definition should be fully understood because the airframe is referred to quite frequently in technical aircraft books.

Because basic airframe weight varies directly with the design loads selected, our real interest lies in determining the maximum value of the design limit load factor—that is, the maximum actual loading the airplane will experience during any anticipated flight condition. This depends largely on the purpose for which the airplane is being designed, with minimum applied load factor values established by FAA design requirements. For example, pleasure and business aircraft are designed for normal category operation at a 3.8-g limit load, while an acrobatic category airplane must be capable of stunt flying at a 6-g limit. Utility category aircraft intended for rugged, bush-type operations are stressed for 4.4-g limit load conditions, with high-performance sailplanes designed for 5.33-g limit load factors. Restricted category aircraft meet at least a 4.2-g limit.[3]

The applied loads are, in turn, distributed into the skins, ribs, and beams (or framing) of the airplane structure through a rather complex process of detailed calculations. So you can see that the size and thickness of various wing, tail, fuselage, and landing gear members, including exterior skins of all-metal aircraft, are determined by the stress levels developed from Part 23 applied loads specifications.

Because an airplane is held together by bolts, rivets, and structural members designed by these applied loads, it is important that they be correctly calculated for the intended use. Once flight limits have been established, the airplane can be placarded not to exceed certain design speed and weight limitations. But just what conditions create structural loads and how can operational limits be applied to aircraft design?

Aircraft loads originate during two distinctly different operating conditions, flight and landing. A special combination of these loadings, the *crash load condition*, is now receiving deserved attention, and is further discussed in chapter 12. Because landing loads are covered for both land and water operation by chapters 14 and 15, the most important set of load conditions, the flight loads, is examined first.

Maneuvering loads

It so happens that flight loads are also of two types: maneuvering loads and gust loads. The word *maneuvering* does not necessarily imply aerobatic flight because such routine actions as a banked turn or a stall above landing speed are considered maneuvers that subjects the airplane to loads greater than 1 g. In level, trimmed, steady cruise flight all parts of the airplane and its contents are subject to a gravitational loading of 1 g. If you weigh 170 lbs on your bathroom scale, you also exert 170 lbs on your airplane in level flight. This loading is carried into the airframe through the seat and floor structure.

When you maneuver the airplane into a 2-g banked turn, your 170-lb body loads the seat and supporting structure at 340 lbs instead of the original 170 lbs, since 2 g × 170 = 340. In similar fashion, the turning maneuver that doubles your body load also doubles the load applied to the wing and other parts of the airplane. Due to centrifugal action, greater lift is required to keep the airplane at a constant angle of bank and also at constant altitude, so a g loading develops. As shown by Fig. 2-1, for a banked turn the limit maneuvering load factor (which is the actual load on the airplane wings divided by the airplane weight at 1 g) increases with the angle of bank Θ in accordance with the relationship $F = W/\text{cosine } \Theta$. The force holding the airplane into a perfect turn is also a function of the bank angle and is given by the equation $L_H = F \text{ sine } \Theta$.

Note that the horizontal force L_H is greater than the centrifugal force if it is caused by an uncoordinated turn. The airplane moves into the turn in a slip. If the horizontal component is less than the centrifugal force, the airplane moves out of the turn in a skid because there is not sufficient horizontal force L_H to satisfy balanced centrifugal load requirements. This is the analysis of how an intentional slip or skid can be developed.

Incidentally, for minimum drag in a turn all controls must be coordinated; as a visual presentation, the ball of the turn and bank (which is the slip or skid) indicator will be centered. For a direct and cheaper approach to eliminating the slip/skid problem, a yaw string can be mounted near the base of the windshield directly ahead of the line of sight. Although some adjustment in string location might be necessary to allow for airflow around the windshield, this can be checked in trimmed, level flight to assure that the string aligns with the centerline of the airplane.

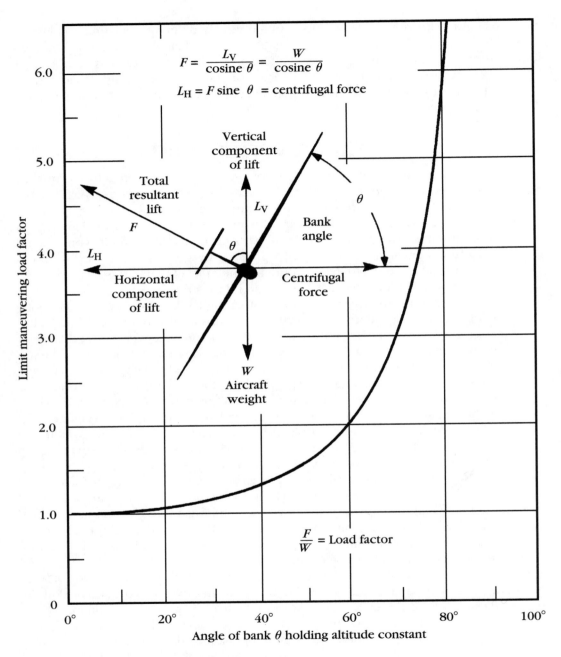

$$F = \frac{L_V}{\text{cosine } \theta} = \frac{W}{\text{cosine } \theta}$$

$L_H = F \text{ sine } \theta = \text{centrifugal force}$

$\frac{F}{W} = \text{Load factor}$

2-1 *Variation of load factor with bank angle.*

In a slip or skid the yaw string moves away from the direction of the slip or skid, so "centering the string" produces a coordinated maneuver. You might have noted that to minimize drag most sailplane pilots have a yaw string located directly in their line of vision, making yawed flight (a slip or skid condition) immediately apparent. To show that not everyone flies the same, many sailplane pilots believe they experience less sink when carrying some slip in steeply banked turns required for thermalling. Possibly this is so because of some induced body lift.

The limit maneuvering load factor represents the actual load increase experienced by the airplane due to any flight maneuver. Basically, this is represented by the increase in wing-structure loading compared with steady 1-g flight. In our example of a 2-g turn, the wing structure of a 2000-lb airplane would be loaded to 4000 lbs. The limit load factor is then said to be 2. Referring to Fig. 2-1 again, a 60° bank angle induces a 2-g limit load factor. Since the cosine of 60° is 0.50, by simple division $F = W/\text{cosine }\Theta$ becomes $F = W/0.50 = 2W$. Since twice the weight of the airplane is 2 g, the 60° bank angle limit maneuvering load factor of Fig. 2-1 is correct.

As you realize from experience, flight loads can develop from many maneuvers in addition to turning flight. For example, elevator displacement creates a load on the horizontal tail which, in turn, loads the fuselage structure. Elevator action also changes the longitudinal trim, or pitch attitude, of the airplane. This action in turn changes the angle of attack of the wing, and so varies the wing loading. Rudder deflection loads the fin and fuselage structure in side bending and torsion. A combination of aileron, elevator, and rudder action loads everything at the same time; so structural analysis of the airframe can become quite complex.

Acrobatics, of course, load the airplane from propeller to tail light, particularly if not properly performed. Biplanes are popular for aerobatic work because their short span reduces damping in roll, which increases rate of roll for all maneuvers.

Four-wing panels offer the opportunity to mount four ailerons to further increase roll rate. And the biplane is also popular because the drag associated with brace wires and struts, combined with the lower structural weight possible from a biplane configuration, prevents a rapid buildup in speed between maneuvers or during sloppily performed maneuvers.

Since maneuvering load factors vary with the square of the speed ratio because lift is a function of velocity squared (V^2), the inherent speed-limiting capability of the biplane is an essential safety ingredient for aerobatic flight. For example, if you stall at 60 mph in smooth air, the load factor is 1 g—that is, when the airplane stops flying it has no external forces acting on it and so will free-fall similar to a steel ball. At a speed of 120 mph, twice the basic stalling speed for this example, the load factor developed at stall is 2^2 or 4 g. Applied to actual flying conditions this means that when caught in extremely turbulent air, simply reducing cruise speed a bit below twice the stall speed prevents your airframe from being loaded at more than 4 g—regardless of the attitudes imposed on the airplane. However, in extreme turbulence it is possible to experience a 4-g stall and spin-in due to maneuver loads arising from corrective control action, or to experience vertical gusts of such intensity that a wing panel can literally be sheared off.

So regardless of speed/load considerations, avoid thunderstorms whenever possible; and if caught, reduce speed to twice the stall speed and do not overcontrol to maintain attitude or altitude for the reasons just discussed.

In similar fashion, high-performance sailplanes can experience extreme maneuver loadings in the hands of an inexperienced pilot. In addition to indicating that a sailplane will go 40 ft along the glidepath for each foot of altitude lost, a 40-to-1 glide ratio also means that an 800-lb sailplane has only 20 lbs of drag. It takes very little diversion of attention and less time for inexperienced pilots to find themselves moving at 120 mph out of 50 mph thermalling flight. If the sailplane stalls at 40 mph, any sudden pull back on the stick to reduce flight speed rapidly loads the wings to 9 g ($120/40 = 3$, and $3^2 = 9$ g). The result—bent wings at least, and possibly none at all.

The important things to remember about maneuvering loads are:
- They arise from control action.
- They increase with the square of the speed/stall ratio.
- They act on everything in the airplane—airframe, passengers, engine installation, battery mount, baggage, etc.

To graphically establish previously mentioned structural design and flight placard speed limits, the V-n, or the V-g diagram of Fig. 2-2 has been developed. Similar diagrams are included in many airplane flight manuals. Since aircraft terminology refers to load factors as n values to prevent confusion with equations containing the gravitational constant g (of 32.2 ft/sec²), load factors throughout this book are referred to as n values to agree with FAA and industry use. Also remember that in our review the term limit load is used to distinguish between the actual applied load required by regulation (or obtained by flight measurement) and the 50% greater ultimate load value used for detailed design of each structural component.

One of the few design classifications permitting legal flight at an overweight condition is the restricted category, so designated by the FAA because operation of this class of airplane must be for specialized purposes and under a very restrictive license. When overloaded, such aircraft are using some of that 50% margin of safety designed into every airplane. As a result, restricted category aircraft are frequently denied the right to fly over populated areas or to carry passengers. Agricultural spraying and dusting aircraft are generally licensed in the restricted category under FAR 21.25 (refer to *Civil Air Manual 8*) and might be permitted to fly at a weight above licensed gross weight in accordance with Fig. 2-3 (taken from CAM 8, Fig. 7-1).

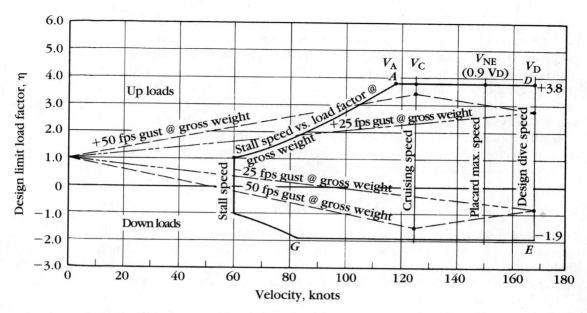

2-2 *The V-n diagram for a normal category medium performance airplane.*

2-3 *Restricted category permissible weight increase.*

It is not advisable to fly at more than design gross weight in private aircraft, and any mention of CAM 8 overweight provisions should not be construed to suggest such action. Agricultural (Ag) pilots and other restricted-category aircraft pilots are usually experienced professionals who fly continually, have excellent control coordination, and risk no one's neck but their own. Whether designing or flying a Part 23 normal category airplane, remember that the 1.50 factor of safety has been included to cover special conditions such as locally marginal material properties and assembly, sudden maneuvers, mishandling in flight, severe turbulence, and the effects of old age—but not intentional overloading, which also reduces climb performance and flight safety.

Gust loads

Despite the 1.50 structural safety margin, properly handled aircraft occasionally experience airframe failures from what are known as gust loads, our second flight load condition. Gust loads arise from the action of turbulent air upon the airplane and might be induced by vertical or horizontal air mass movement.

We are all familiar with thermals that produce vertical gusts of a generally minor nature, although they are representative of the type of flow producing major gust loads upon the airplane. Frontal action preceding a storm, vertical air motion resulting from winds blowing at right angles to a line of hills (ridge lift), clear air turbulence at altitude, and wind rotors associated with lenticular cloud formations are all capable of producing severe and sharp structural loads. FAA design requirements have recently been revised to consider more fully gust effects on the airframe, and now require the structure to withstand 50 ft/sec gusts at cruise and 25 ft/sec gusts at maximum design speed.

When you consider that a 50 ft/sec vertical gust represents the impact of an air column rising at 3000 ft/min (34 mph), the energy contained in gust action becomes more apparent. Fortunately, private aircraft are generally designed with comparatively light wing loadings of about 14 lb/sq ft and so ride gusts in a manner similar to a cork on a wave crest. However, business and commercial jets experience sharp gust loads due to higher wing loading and greater airplane inertia; so jet passengers are more comfortable, but the airframe is not.

Gusts can also act in a horizontal direction and can be particularly apparent and annoying during a landing in frontal passage conditions. Wind shear, a comparatively new term in our flight weather vocabulary, raises its head in these circumstances and is usually overcome by approaching at higher than normal speeds. Added speed means safety when the horizontal gusts range 20 to 30 mph above the average wind speed, since any sudden drop in gust velocity could stall an airplane approaching at normal speeds. For example, if stall occurs at 60 mph and the approach is at 80 mph with surface winds gusting to 45 mph, a sudden 20 mph drop in headwind velocity back to 25 mph causes the airplane to stall (80 mph indicated approach speed minus 20 mph drop in headwind velocity equal 60 mph stall speed).

This horizontal gust effect frequently occurs near the ground and can drastically affect approach glidepath angle as well as stall margin. If an approach is established along a 4° glidepath, a 20 mph headwind increase steepens the angle to about 7°, requiring corrective action if the gust continues for any period of time. Under this set of conditions the tendency is to raise the nose, reducing airspeed and descent rate, but heading directly into a stall/spin condition if the gust suddenly terminates. Unfortunately, this situation is quite typical of the approach accident. For safe operation, plan to increase your normal approach speed during gust conditions by an amount equal to the reported gust intensity. For example, if runway winds are reported at 10 mph gusting to 20 mph, increase your approach speed by 10 knots.

So you see that in addition to loading the aircraft structure, horizontal and vertical gusts also affect flight handling and safety. These conditions can exist at altitude as well as near the surface, although gust intensity is usually lower above 20,000 ft.

The flight envelope

Now that maneuvering and gust loads have been described, let us return to the diagram of Fig. 2-2. This is a typical flight envelope for a moderate cruising speed general aviation airplane. The maximum positive limit load factor is +3.8 g running from the stalling speed at the load factor, Point A, to the design dive speed V_D of Point D.

The value of V_D is established by FAA requirements based on the design cruising speed V_C. In accordance with FAR 23.335, V_C must be established as a function of the wing loading, unless a lower value of V_C can be substantiated by aerodynamic calculation or flight test demonstration. For normal category aircraft FAA criteria set $V_D = 1.40\ V_C$; for utility category operation, $V_D = 1.50\ V_C$; and acrobatic aircraft are designed with $V_D = 1.55\ V_C$. These are important basic design speed values because V_C and V_D are used to obtain wing structure loadings, while the airplane must be demonstrated to be free from flutter at V_D.

The never exceed speed, which is the placard red line speed, V_{NE}, is established at 0.9 V_D by FAA design requirements. As a safety feature, every production airplane must be factory dive-tested to V_D, which is a speed 10% greater than the owner can legally fly.

Gust load conditions are shown by the dashed lines running to cruise speed V_C, and then to V_D. While gust conditions are not crucial for the wing structure of the airplane represented by Fig. 2-2 because their value is below +3.8 load factor, gust loads probably will design the horizontal and vertical tail surfaces as they do for most modern aircraft. Tail loading will be reviewed at greater length in chapter 7.

The negative loadings of Fig. 2-2 are half the positive values, indicating this diagram applies to a nonacrobatic, normal category airplane. This type of loading is particularly important to strut-braced high-wing monoplanes, which have the strut placed in compression by any down load on the wings. If the down loads were as great as the up loads, the support strut could be of considerable size and weight.

Fortunately, flight load investigations have indicated the range of +3.8 to −1.9 limit load factors to be safe for normal aircraft operation. However, for simplification of wing beam construction plus a bit of extra structural safety, cantilever monoplanes can be designed at slight additional weight to have equal strength under up or down wing load conditions; that is, wings capable of +3.8 or −3.8 limit load factors. This feature should be considered for amphibian aircraft equipped with wing floats, since floats transfer repeated water impact and structural rebound fatigue loads directly into the wing structure.

While load and speed values vary from one airplane to another, the flight envelope diagram of Fig. 2-2 is representative for all modern aircraft and is the basis for airframe structural calculations, since it documents the design load factors at all permissible flight speeds and weights.

Level flight loads

The airplane is loaded not only during maneuvering and gust conditions but in level flight as well. When all loads are in balance, level flight is maintained at constant airspeed and altitude. The forces acting in trimmed flight are shown in Fig. 2-4 for two different thrust line locations.

Most aviation texts refer to four forces of flight: thrust, drag, lift, and weight. However, there is a fifth force that interacts with the basic four—this is the horizontal tail balancing load represented by the symbol B of Fig. 2-4. This value must be carried by the wings when the tail load is down, since both weight W and tail load B must be balanced by wing lift L if the airplane is to maintain level flight at constant altitude.

Obviously, the lower the tail load the lower the total wing lift required; and less wing lift means less wing drag, which results in more cruising speed. As a result, the search for reduced horizontal tail loads has produced lifting aft surfaces (the Flying Flea and similar tandem designs), the canard configuration (which is equipped with a lifting tail located ahead of the wings as shown in Fig. 3-4), and three surface aircraft.

Most modern aircraft develop relatively low horizontal tail loads by using main wing airfoils having low section moment coefficient values. The section moment coefficient C_{Mac} represents the summation of all aerodynamic forces acting upon the airfoil. These aerodynamic forces are mainly the components of section lift and drag pressures resulting from airfoil surface curvature, and the surface friction drag unbalance between the upper and lower surfaces acting as an aerodynamic couple.

Figure 2-5 shows why the wing section moment M_{ac} normally acts to rotate a wing leading-edge down (creating a counterclockwise moment). The wing

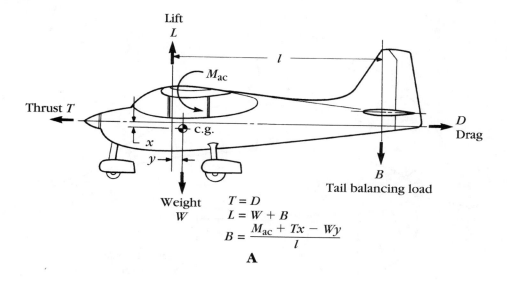

$$T = D$$
$$L = W + B$$
$$B = \frac{M_{ac} + Tx - Wy}{l}$$

A

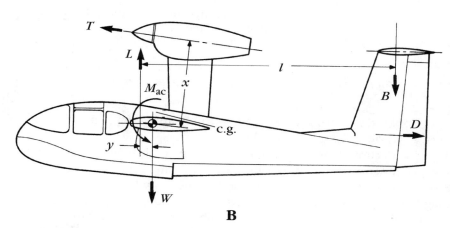

B

2-4 *Level flight loads. (a) Forces acting upon the normal private airplane; trimmed cruising flight condition. (b) Forces acting with an offset thrust line.*

section moment acts about the aerodynamic center of the airfoil selected, the aerodynamic center being the locus of all external aerodynamic forces acting on the wing section as shown in Fig. 2-5. The aerodynamic center differs slightly in location from the airfoil center of pressure because the center of pressure is determined by summation of only the pressure forces acting on the wing. As such, the center of pressure is an important indicator of lift movement along the wing chord with variation in angle of attack and airspeed, while forces acting about the aerodynamic center create the torsional loads acting on the wing structure.

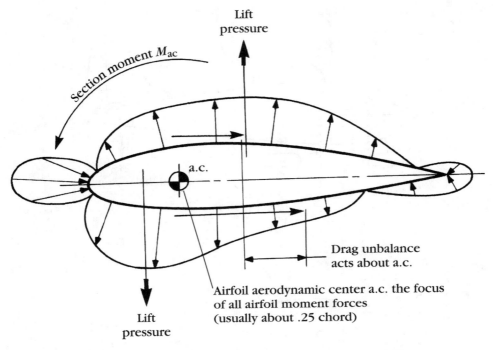

Lift
pressure

Section moment M_{ac}

a.c.

Drag unbalance
acts about a.c.

Airfoil aerodynamic center a.c. the focus
of all airfoil moment forces
(usually about .25 chord)

Lift
pressure

2-5 *Airfoil section forces produce counterclockwise wing torsional moments.*

For most applications, both centers can be safely approximated at the same location of 25% wing chord aft of the leading edge.

M_{ac} is a function of section moment coefficient, air density, wing area, airspeed squared, and wing chord; it increases with any one of these factors. Because the wing section moment tries to rotate the wing leading edge down, the
resulting twisting load acts on the wing structure and the wing-to-fuselage connection to provide the torsional design loading for this important joint. In order
to keep the fuselage as well as the wing from nosing down, a balancing force
must be developed, such as the horizontal tail load B of Fig. 2-4. So the wing
moment value M_{ac} has an important effect on the entire airplane design.

Figure 2-4b shows the thrust line position with respect to the airplane center of gravity, which also acts to load or unload the horizontal tail. The T tail
tends to be self-balancing for such thrust line locations, however, since tail-
down load increases with slipstream flow to counterbalance the nose-down moment resulting from power increases. Regardless of the degree of self-trimming
obtained, the additional balancing tail load must be supported by the wings if
level flight is to be maintained.

A specialized approach that reduces horizontal tail loads is used in sailplane
design. Because power changes are not a problem, the normal center of gravity

is moved as far aft as flight control requirements will permit, usually to the point of neutral stability where the airplane demands constant pilot attention for controlled flight.

Looking at the diagrams and the B value formula of Fig. 2-4, you can see why the tail load decreases as the value of y increases; lift L acting upward ahead of the airplane's center of gravity tends to overcome the nose-down wing section moment M_{ac}.

For sailplanes, a combination of aft center of gravity plus a long tail arm (l value) results in low tail load and less wing support load, reducing required lift and wing drag values. This load arrangement results in low drag flight and reduced airframe loads, providing an efficient design configuration. Similar load optimization is necessary for all successful aircraft and requires considerable study during the preliminary design phase of every new aircraft program.

Before concentrating on the design features, performance requirements, and handling characteristics peculiar to the single-engine private airplane, some additional design background is most helpful. I'll first review the persistent design variations that never quite make the grade in an endless search for utility, and then move on to a brief discussion of successful configurations that are always popular.

Endnotes

[1] *FAA Federal Air Regulations (FAR) Part 23; Airworthiness Standards: Normal, Utility, and Acrobatic Category Airplanes.* Available for approximately $35 subscription and provides all amendments issued from the Superintendent of Documents, U.S. Government Printing Office, Washington, DC 20402.

[2] "Export Airworthiness Approval Procedures." Appendix 2 of the *FAA Advisory Circular No. 21-2.*

[3] FAR 23.337 for powered aircraft; FAR 21.24 for Primary Category Aircraft and Gliders (also FAA *Basic Glider Criteria Handbook*); and FAR 21.25 for Restricted Category Aircraft (also FAA CAM 8).

3

The search for utility: repeated attempts at progress

With the exception of pilots primarily interested in flying as a relaxing or challenging sport, the private airplane basically appeals to those who can save time or expand their operating range through its use. Both reasons are the same and relate to cruising speed plus the ability to fly more or less in a straight line between terminal points. For such users the private airplane provides a form of transportation not possible with any other vehicle. In addition, for a trip of any distance, the airplane offers less fatiguing travel. Unfortunately, the limit of aircraft utility is reached upon arrival at the destination airport. Any further travel necessary to make a business call or visit friends depends on the availability of a taxi, rental car, or someone to pick you up—all of which require some degree of prior planning. In addition, weather frequently interrupts a flight before it begins or causes en route delays.

In an effort to overcome these drawbacks, the quest for a universal airplane capable of safe off-field landings, with the ability to travel highways, and with ever higher cruising speeds has appealed to aircraft designers from the moment efficient horsepower first became available. However, the pathway to this goal is strewn with hundreds of failures in virtually all possible aircraft configurations. But along the way, a few types did evolve into today's accepted aircraft that will be reviewed in chapter 4. Although we still lack the ultimate universal airplane, why have some designs survived while others have been rapidly rejected?

The answer apparently lies in some combination of the degree of skill required to pilot the airplane, the utility offered per dollar of initial cost, and the difficulty of obtaining FAA certification approval. Not surprisingly, a few basic design types seem to lead the list of repeated attempts to improve utility. By way of broadening your aviation background and with some slight hope of furthering progress, I'll briefly review the historically important, recurring "unconventional" configurations. Quite possibly a discussion of past problems will favorably influence future development.

The autogyro

The most persistent design is that of the autogyro. This type of rotating wing airplane was even popular back in the 1930s when autogyro mail service was inaugurated in the Philadelphia area, then the center of rotor aircraft activity. The autogyro of Fig. 3-1 was developed into a roadable design that actually flew and was frequently driven on main highways during the mid 1930s.

3-1 *Roadable autogyro developed during the thirties.* Aviation Magazine

Designed by the Autogyro Corporation of America (ACA), this small two-place ship was financed by the Bureau of Air Commerce for flight and road-handling evaluation. Although total production was limited, autogyros were manufactured prior to World War II by Kellett Autogiro and Pennsylvania Aircraft Syndicate (Wilford Gyroplane), as well as ACA. This concept reappeared during the 1960s and 1970s as the two-place Umbaugh (later Air and Space Manufacturing) and J-2 Jovair (later McCulloch) designs. As you can see, the recent models have had difficult production lives even though both received FAA Type Certificates under FAR Part 27. As of this writing, neither autogyro is in active production. However, the J-2 Gyroplane of Fig. 3-2 shows the modern trend in autogyro development.

Jovair Corp.

3-2 *Prototype Jovair J-2 Gyroplane in flight.*

Why does this design persist and why doesn't it find acceptance? The primary attraction is the ability to take off and land with full STOL (short takeoff and landing) performance. This capability can be expanded to achieve essentially vertical takeoff by driving the rotor to greater rpm than obtained in flight, then declutching and applying pitch to the rotor system. This sudden buildup in lift literally snatches the autogyro about 50 ft into the air, when rapid application of forward power and the still lifting rotor combines to provide forward flight. Of course, you might ask why anyone would want the mechanical and control complexities necessary for jump takeoff for a design that does not require rotor power for normal flight. This is a good question; and since not many people need such limited vertical flight capability, the related cost in weight and dollars tends to eliminate this feature of autogyro operation. For somewhat more initial cost, the owner requiring frequent vertical flight can purchase a small helicopter that also has greater useful load.

In addition, the autogyro rotor system is more expensive to purchase and requires more maintenance than fixed wings. Even without a rotor drive system, the controls are complex, while the combined weight of the rotor and pylon exceeds that of fixed wing area for an airplane of comparable gross weight. With somewhat different operational and handling techniques to be learned, thorough pilot checkout is mandatory for safe operation. As a final consideration, the autogyro has a higher empty weight and lower useful load than a fixed wing airplane of comparable power. In fact the 180-hp J-2 useful load is only 510 lbs.

Then what is left to benefit the autogyro owner? Only the fact that the autogyro is capable of flying more slowly than a fixed wing airplane, which permits a steeper angle of climb and lower landing speed. The autogyro is a novel approach that does not offer the utility of a fixed wing landplane of comparable gross weight, power, or price. It has a natural market only among the few owners who must have some STOL capability at minimum price. As such, past autogyro designs had nothing to offer and accordingly have never been successful.

If someone develops a simple rigid rotor autogyro having cruising speed, empty weight, and useful load similar to a fixed wing airplane of equivalent power, acceptance would change rapidly; but such a configuration appears fundamentally impossible.

Roadable aircraft

The attraction of being able to use the same vehicle to travel from home to a distant point is almost as persistent as the autogyro concept. Fostered by pioneer roadable autogyro development, the desire for door-to-door travel averaging more than 100 mph proliferated after World War II and remains alive today. Such companies as Convair, Fulton, Spratt, and Aerocar, plus numerous homebuilders, ventured into various types of roadable airplane development. Of these, only Aerocar ever received FAA certification, and that approval was for a limited production design. While Aerocar, under the direction of Molt Taylor, remains active in the development of roadable aircraft, the automobile is still the only road vehicle available. As would be expected, there are sound reasons why this is so.

In addition to the obvious economy of keeping your airplane at home, one of the great advantages offered by the roadable airplane is the ability to land ahead of approaching bad weather and continue traveling by road. To be effective in combating weather, it is essential that all airplane parts continue the journey. Past designs have hindered roadable development by overlooking this.

To leave wings or wings and tail surfaces behind during ground operation means that the journey has to be retraced to reassemble the various flight components. This design approach is not practical simply because of the time lost in retrieval effort. In addition, if clear weather is reached by driving along the road, flight is not possible with the wings left behind. And, unless housed, the temporarily discarded surfaces are subject to wind damage and vandalism.

The FAA certified Aerocar of Fig. 3-3 has the optional capability of trailing all its components, or leaving the flight structure in the garage or small hangar at the home airport. The Aerocar can be an airplane, a roadable airplane, or simply a compact automobile.

3-3 *Aerocar III. (a) As ground vehicle. (b) Ready for flight.*

Although most of the complex development required to reach this state-of-the-art roadable aircraft is not apparent, the major problems are:

- steering on the road, which is much too sensitive with an aileron-type direct connection to the wheels;
- adequate road suspension using four wheels that can be retracted in flight;
- powering two wheels for road operation;
- developing simply connected wing, tail, and propeller drive attachments; and
- cooling the buried engine.

These features all add up to considerable vehicle weight and manufacturing cost. As a result, the roadable airplane offers superior, in fact, almost ultimate utility, but at increased cost and decreased useful load compared to a conventional airplane of similar gross weight.

This is an appealing trade-off, and one that could encourage broad acceptance of the roadable airplane. Unfortunately, the real problems are more political than technical and begin with the transition from flight to ground operation. For at that point existing FAA flight standards along with federal and state motor vehicle regulations combine to defeat the roadable concept.

Because of weather conditions it might not be possible to land at an airport, presenting the likelihood that FAA airspace regulations will be violated if a road or roadside landing is made. But this prospect is minor compared with the situation faced when reaching the highway. When operating as an automobile, the vehicle must comply fully with state motor vehicle laws. These laws include headlight, safety light, and exhaust emission standards; specified roadway design features; federal and state automobile taxes; liability insurance minimums; etc. To this add the certainty of being arrested in most states for landing on a public highway, to say nothing of local protests and resentments. As Molt Taylor of Aerocar indicates, the political difficulties of roadable airplane operation in our changing society truly outweigh the technical development problems.[1]

Although the design features associated with roadable aircraft development are formidable, they can be solved with existing technology. However, the roadable airplane's practical and legal complications have delayed, if not prevented, the introduction of an improved method of transportation. The FAA could assist in hastening acceptance by acting as a focal point to define and develop roadable aircraft specifications and regulations that have the approval of air traffic control, federal highway development programs (roadside landing strips), state vehicle commissions, and various federal and state tax commissions. The task is difficult, but it must be accomplished before the roadable airplane can be economically produced. In view of this, it is unlikely that roadable aircraft can ever realize volume production—unless supported by the power of a major automotive corporation capable of amending present vehicle legislation.

Canard aircraft

The earliest powered aircraft had horizontal tail surfaces located ahead of the wings. Practically all of the pioneers at one time in their early development employed this configuration, called canard by the French because it resembled a duck in flight. I do not know of any advanced technical reason for this initially universal tail arrangement used by such experimenters as the Wrights, Curtiss, Langley, Avro, Short, or the Farmans. The tail first arrangement seems particularly strange because four of the most technically-sound basic experimenters, Lilienthal, Chanute, Montgomery, and Pilcher, used a (now conventional) aft location for the horizontal and vertical tail surfaces of their 1890s gliders. Possibly the powered designers copied the Wrights' successful configuration, and then found this horizontal tail location a convenient reference for coordinating surface deflection with longitudinal trim response.

Still, the concept of a lifting forward tail is an appealing way to carry some of the airplane weight and help reduce drag and increase cruising speed. Conventional airplane designs that have tail surfaces located behind the wing use the horizontal tail to balance flight, thrust, and center-of-gravity travel loads. This usually means a down load on the tail, as discussed in chapter 2 and shown in Fig. 2-4, requiring an increase in lift coefficient to support the added wing load. Since wing drag increases with wing lift, a climb and cruise penalty is paid for the stability offered by an aft-located horizontal tail.

As shown in Fig. 3-4, a forward location of the horizontal tail divides the lift between two surfaces regardless of the load condition. Comparison with the bal-

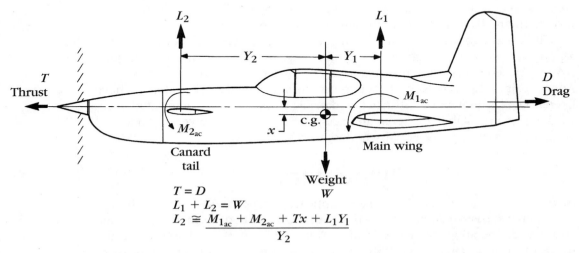

$$T = D$$
$$L_1 + L_2 = W$$
$$L_2 \cong \frac{M_{1_{ac}} + M_{2_{ac}} + Tx + L_1 Y_1}{Y_2}$$

3-4 *Balancing forces acting upon a canard (horizontal tail forward) aircraft configuration.*

ance formula for tail load B of Fig. 2-4 clearly shows that determining the canard load L_2 is a more complex and crucial problem than it is for the conventional aft-tail location.

The canard arrangement offers the safety of a virtually spinproof airplane because the forward wing is designed to (gently) stall before the main wing can stall. At forward wing stall, the lift and section moment of the main wing acting aft of the airplane's center of gravity reduces the pitch trim angle until lift is restored to the forward wing, usually with minimum loss of altitude. This stall relationship is a mandatory design criterion for the canard configuration and must be combined with the airplane's center of gravity located forward of the resulting center of lift of both wings.

When the main wing has flaps lowered, the increased longitudinal nose-down moment is balanced by added lift supplied by the forward (canard) wing. This trim lift can be realized either by increasing the canard angle of incidence, using elevators on the canard, or both. By increasing the angle of incidence of the forward wing (wing chord line angle relative to the fuselage reference line), the required trim lift can be realized without increasing the fuselage longitudinal trim angle. Of course, a greater trim angle increases the drag of the canard surface, but not to the degree resulting from an increase in fuselage trim angle.

Because trimming the entire forward surface is more complex than lowering forward elevators to increase canard lift, elevator trim is usually preferred. Either way, forward wing lift available up to the stall of this surface imposes a trim limitation on main wing flap deflection (reducing maximum lift coefficient) and forward center of gravity position. While this restriction of main wing lift to prevent main wing stall results in an increase in minimum flight speed, as well as increased takeoff and landing runs, the favorable trade-off can result in a spinproof airplane having efficient cruising performance.

The canard configuration has been successfully revived and developed into a popular homebuilt design by Burt Rutan. Shown in Fig. 3-5, the VariEze set a world's closed-course distance record at the Experimental Aircraft Association Annual Convention in Oshkosh, Wisconsin during August 1975, and many have been flown by homebuilders since that date. For those seriously interested in canard design and stability criteria, an excellent basic paper is given in endnote,[2] while Rutan's work is covered by endnote.[3]

Flying wings

At one time, new flying wing designs appeared in powered form on about a semiannual basis; their mission was to reduce drag and so increase cruising speed and operating efficiency. Waldo Waterman's work probably came the nearest to private aircraft certification, while Northrop Aircraft carried this concept to the ultimate with the 172-ft span, 400-mph XB-35 built at the end of

Rutan Aircraft, photo by Dan Dwiggens

3-5 *The canard VariEze in flight.*

World War II. Repowered with jet engines, this ship became the YB-49 flying wing of 1950, but it never reached production.

The Vought XF5U-1, designed by Charles Zimmermann of NACA (National Advisory Committee for Aeronautics), was a particularly interesting version of the flying wing configuration in that the wing was essentially of double elliptical shape and of very low aspect ratio. I had the pleasure of working on the twin-propeller powerplant drive system for this design in 1942. Although having sufficient thrust to hover in flight (while looking like a pumpkin seed on edge), and successfully flown in prototype form, the XF5U-1 was abandoned when jet-powered fighters arrived with flight speeds beyond the capability of this maneuverable design.

The basic difficulty with the flying wing configuration is stability, or actually, the lack of stability about all three axes—pitch, roll, and yaw. Although very sophisticated autopilot controls were designed into the Northrop military wings, such devices exceed the cost and maintainability levels acceptable for private or even commercial aircraft.

The flying wing configuration is kept alive today by the Air Force in the form of the Northrop B-2 bomber and by private experimenters working with gliders, Rogallo wings, and small powered aircraft, although current FAA stability requirements preclude type certification of any simple design, and so this concept might not be suitable for private aircraft production.

Three-surface aircraft

The original Curtiss, Farman, Voisin, Cody, and similar aircraft used a three-surface arrangement. Most of these configurations had a fixed set of aft surfaces, with longitudinal trim provided by an all-moving horizontal tail located ahead of the pilot. In 1911, Graham-White added an aft elevator modification that moved in coordination with the forward elevator surface.

When working on the Curtiss June Bug replica for the Glenn H. Curtiss Museum in Hammondsport, New York, and more recently on a 1910 Farman biplane flying replica for the Owls Head Transportation Museum of Owls Head (Rockland), Maine, it became quite necessary to determine what the original designer intended the forward and aft horizontal surfaces to do. In each case I assumed that the aft, fixed horizontal tail surfaces were used to provide lift, while the forward horizontal surface was intended to operate as a longitudinal trim elevator, positioned where surface displacement could be clearly observed by the pilot.

This three-surface arrangement permitted a comparatively large range of center of gravity (c.g.) travel depending on the fixed incidence adjustment of the aft horizontal surfaces. This permitted the early experimenters to allow for (possibly unexpected) c.g. variations between various production aircraft of similar design.

This concept is in vogue once again on such aircraft as the Piaggio P.180 Avanti, experimental Rutan configurations, and a few new high-performance aircraft under development. Depending on the relative size of each horizontal surface, the three-surface arrangement permits locating the c.g. virtually where desired, even to the extent of balancing an aft located tractor powerplant. However, the interaction between the forward and aft surfaces to provide longitudinal trim can be extremely crucial, not only to develop acceptable low-speed flight without overloading the forward surface, but also to assure that the forward surface will stall before the aft and center surfaces in order to provide a positive (nose-down) recovery attitude.

Fortunately, modern computer techniques such as computational fluid dynamics (CFD) and finite element analysis (FEA) permit detailed study of surface interaction and incidence settings. This detailed study readily shows the effects of surface area, airfoil section, trim surface deflection, and wing incidence upon airflow, airplane lift coefficient, and c.g. location. As a result of being able to tailor all surfaces to provide low-drag lift during cruise flight, you can expect to see a number of three-surface designs appear in the near future.

VTOL aircraft

If a simple configuration could be developed, the VTOL (vertical takeoff and landing) airplane would be the ultimate flying machine. The ability to take off and land vertically with present aircraft-type control, powerplant, and propeller systems would provide maximum utility and safety. Imagine being able to come to a midair stop when encountering unexpected IFR (Instrument Flight Rules) conditions and then slowly descending to a convenient landing site until the weather improved.

The VSTOL's appeal is quite universal. Although the helicopter does all of this, it is expensive to purchase and to maintain, requires much greater pilot skill than fixed wing aircraft, and is not suitable for IFR operation unless very expensively equipped. As a result, there have been repeated attempts at some version of fixed wing VTOL flight.

I became interested in developing a personal airplane VTOL configuration in the mid 1970s. I built a flying model powered by rather sophisticated, small electric motors. Developing flight problems and further analysis clearly indicated that perfection of the design, to say nothing of FAA certification, was well beyond the design capability and pocketbook level of any single individual.

The control and trim requirements during hover are severe, due to the rigid hub propellers, and particularly in gusty wind conditions, so the control system tends to become complex. Quite possibly, if we could be content to land in a 100 yd sq area instead of 100 ft sq area, the degree of control sophistication required would be reduced.

Stability is a problem, however. The stability problems associated with fixed wing VTOL operation exceed those of the flying wing, being particularly complex during the transition from vertical lift to forward flight and vice versa.

Today, the Bell Boeing V-22 is in final stages of development, although that path has been traveled at great human and financial cost. The noted stability problems have been generally overcome. While the concept remains attractive, it requires great simplification before a similar VTOL aircraft configuration can be marketed for personal use. But because this concept offers the most utility and safety of any airplane configuration, we might reasonably expect acceptable fixed wing VTOL aircraft to be developed within the next 25 years.

Endnotes

[1]M.B.Taylor, "The Aerocar Flying Automobile." *Society of Automotive Engineers Report 740393*, April 1974.

[2]Joseph V. Foa, "Proportioning a Canard Airplane for Longitudinal Stability and Safety Against Stall." *Journal of the Aeronautical Sciences*, December 1942, pp. 523-528.

[3]Burt Rutan, Rutan Aircraft Factory, Bldg. 13, Mojave Airport, P. O. Box 656, Mojave, California 93501.

4

Why that configuration?

Like automobiles, aircraft designed for the same use are becoming so similar in appearance that it is extremely difficult for even an experienced observer to tell them apart. Of course, the high-wing designs can be separated from the low-wing models, but try to distinguish a Beech Musketeer from a Piper Cherokee in the landing pattern. And even all the high-wing Cessnas look much alike because they were derived from the same basic design.

While anyone interested in flying can readily tell a sailplane from a powered airplane, it takes an owner to separate and identify the various models of high-performance sailplanes. Another example of configuration similarity is the small biplane, which dominates the acrobatic field because of its ability to perform complex maneuvers. It becomes apparent that aircraft designed for similar use look similar because of the performance and structural demands peculiar to the service required.

As indicated by the title, the real purpose of this book is to analyze and describe how various design features and requirements affect aircraft performance and mission capability. Incidentally, the terms *private* or *personal aircraft* refer to single-engine pleasure, business, and training planes, a type basic to the general aviation industry and familiar to every pilot.

In order to understand more fully the effect of design on performance, I'll briefly analyze the characteristics of major types of aircraft in current use. Private aircraft design features are covered in chapter 1, so I won't review them again. However, I'll consider the different and special criteria for business twins, agricultural, acrobatic and sport-flying aircraft, sailplanes, and seaplanes.

Following this chapter, just how operational requirements actually mold the final aircraft design is discussed, including the various options in the owner's manual or aircraft flight manual style. Just how and why design requirements result in specific airplane configurations tailored to the intended mission will become increasingly apparent as you review these different aircraft.

Business twins

The less than 6000-lb gross weight twin-engine airplane is one of the most useful types of planes available for the general aviation market. Since such aircraft are primarily used to go someplace on schedule if legally possible, the design must perform acceptably under IFR conditions. Reliability and safety are important requirements, so two engines are quite logical, provided the airplane can maintain level flight or climb slightly on one engine. Icing protection is also necessary for practical operation.

Because two engines are used to assure flight safety in the event one fails, the crucial design consideration will be ease of handling with one engine out. Stability under such adverse conditions becomes important, so a relatively long tail arm is necessary. A large fin is also required to control the yawing moment produced by the combination of offset thrust from the operating engine plus drag from the dead engine.

Another important design requirement for the modern business twin is a high cruising speed. To realize high cruise speed, a low total power loading approaching 10 lb/hp plus a clean aerodynamic design are necessary and both are mandatory to obtain acceptable single-engine operation. Turbocharged engines make a somewhat more complicated powerplant installation, but they provide altitude power for improved cruise performance as well as a greater margin of power for single-engine operation at intermediate altitudes.

This type of airplane must have a range of 600 to 1000 statute miles in order to cover intermediate trip distances faster than commercial transports. Endurance is also necessary to meet alternative destination requirements when operating in actual IFR conditions, so considerable fuel capacity must be provided to feed both engines.

For passenger safety and reduction of wing bending loads, all fuel should be stored in the wings. Naturally, a mass of fuel distributed along the wing span and/or at the wing tips reduces the rate of roll. As a result, the twin must have powerful ailerons to satisfy crosswind landing requirements when loaded, a consideration demonstrated by Piper when they increased the aileron span for their 1976 model Seneca II of Fig. 4-1.

The requirement for high cruising speed results in reducing wing area to lower total drag, which in turn produces higher landing and stall speeds. This means powerful flaps are necessary. Although high-wing loading tends to reduce climb rate and ceiling, any such loss is more than replaced by the increased climb power available from two engines plus low configuration drag.

Business twins are generally used as personal transport aircraft carrying four to six passengers, making adequate ventilation and superior soundproofing basic design considerations. Ground ventilation in particular can be crucial, so it is not surprising that cabin air conditioning systems have been developed as optional equipment for modern twins. Due to IFR flight communication and

Piper Aircraft

4-1 *Piper Seneca II.*

navigation requirements, the instrument panel must have space for plenty of electronic equipment and flight instruments. Of course, FAA airway and traffic control regulations dictate the need for a great deal of electronics and instruments, which are quite common for all aircraft and certainly not limited to business twins.

Because twin-engine airplanes are usually flown by experienced pilots who fly more frequently than the single-engine rated private pilot, stall characteristics are not as crucial as for private aircraft. In fact, a gentle or prolonged stall peak is of secondary importance to minimum section drag and moment characteristics, with modern designs using a flush-riveted laminar-flow airfoil wing.

Business twins generally operate from paved runways and at comparatively high takeoff and landing speeds; so retractable tricycle gear is used for ease of ground handling, improved flight safety in crosswind conditions, and minimum cruise drag. However, for rough terrain and bush-country use, consideration should be given to tail-wheel-type landing gear, as will be discussed in chapters 6 and 14.

Although there are not many practical locations for just one engine, twins do have two power packages that can be moved about. Cessna designers have developed a tandem-powered passenger pod locating the thrust of both engines on the aircraft's center line. As shown by Fig. 4-2, the Cessna 337 design does not experience yawed flight from offset thrust when either engine fails, although some flow change might require minor trim corrections. Since less yaw means less drag with the same power, this design can maintain a higher single-engine ceiling at gross weight than is possible with engines lo-

4-2 *Twin-engine Cessna 337 Skymaster.*

cated off the centerline. Or conversely, under single-engine operation this configuration can carry greater weight to a given altitude than can an offset twin design of similar power.

Of course, designers have worked on other methods to reduce single-engine drag, not only to improve performance but also to reduce the cabin workload. One of the first and primary methods was the elimination of windmilling prop drag by feathering a dead engine propeller (so the blade section parallels the airflow). Opposite rotating propellers are sometimes used to balance airflow over the wing and tail, reducing trim changes with power variations for both twin-engine and one-engine-out conditions. The feathering propeller remains the most productive aid currently available for single-engine operation. Successful business twins have the following common features:

- high cruising speed and twin-engine range;
- IFR capability with icing protection;
- excellent stability plus comparative ease of handling with one engine stopped;
- relatively high power for climb and cruise performance as well as single-engine operation;
- long tail arm and large vertical tail for directional stability and single-engine control;
- low-wing area compared with private single-engine aircraft plus a low-drag airfoil section;
- long-range cruise fuel capacity with fuel preferably located in the wings;
- powerful ailerons and flaps;
- feathering propellers, possibly counter-rotating;
- tricycle-type landing gear; and
- superior cabin soundproofing with ground ventilation.

Although all production twins seem to follow this design format, I believe a market exists for owners who want two engines primarily for cross-country and IFR safety, rather than for great speed or altitude performance. Since they would only occasionally carry four people, this design could have a Cherokee 2 + 2 type cabin, two small 100- to 125-hp engines, constant-speed propellers, fixed faired tricycle landing gear, and about a 600-mile maximum range. Feathered props should not be necessary because single-engine yaw forces would be low; elimination of feathering along with the fixed gear will reduce purchase and maintenance costs as well as cockpit workload. What more could the occasional pilot ask for?

Such a design could carry two passengers plus much baggage and fuel at 150 mph safely over rough mountains and through nonicing IFR conditions, day or night—and at half the cost of present twin-engine designs. Quite possibly the present twin-engine design configuration needs more thought and some revision.

Agricultural aircraft

We now turn to a very specialized type of airplane design; actually, a valuable farm tool and one capable of repaying its purchase cost within one to two years—the agricultural (Ag) airplane. Since the prime function of this type of aircraft is to spray, dust, or seed from very low altitude and over practically any type of surface, it is essential that the airplane be extremely maneuverable and have an instantly responsive engine. Maneuverability is mandatory for turning in short fields and around or over obstacles, while instantaneous power response is necessary for sharp pull-ups around these same obstructions or to clear rising terrain.

In the hands of an experienced pilot, the less stable an airplane the more maneuverable it is, at least up to the point of neutral stability. Neutral stability occurs when the airplane continues flying stick free in whatever attitude it has been placed; the stick forces per g are very low, and there might be no increase in stick load with elevator displacement or increase in rudder pedal load with rudder deflection. More simply, the airplane does not tend to return to its original pitch, roll, or yaw attitude when it has been displaced.

As a result, aircraft with neutral or slight positive stability do not fight the pilot to return to their original flight path, so they are easily and rapidly controlled in pitch and yaw, but they must be constantly flown. As a matter of interest, neutral or slight longitudinal stability is usually associated with a fairly short tail arm. Due to the high degree of maneuverability required, slightly positive (marginal) stability is a basic design parameter for successful agricultural aircraft and is one reason this type should only be operated by trained professional pilots.

Roll rate must also be high for the maneuverability required. Biplane wings are one solution offering plenty of area in a small span, plus the opportunity to mount four ailerons. The short span reduces roll damping while two sets of ailerons increase roll power. Ag planes intended for small, enclosed-area operations such as rice-field control are biplanes because of their maneuverability.

One of the best Ag planes is the Schweizer Ag Cat of Fig. 4-3, which has been in production since 1958 with more than 1800 aircraft delivered to date. Note that this design is quite short-coupled (short tail arm) and has plenty of wing area. In addition to providing a high rate of roll as well as great amounts of wing area, a biplane wing structure is lighter than a cantilever (unsupported) monoplane wing of similar area. Structural weight saved by the biplane configuration increases payload and climb, or can be used to provide a more rugged and crash-resistant fuselage structure.

The Piper PA-36 Pawnee Brave of Fig. 4-4 is noted to have a cantilever monoplane wing representing another design approach. This Ag plane configuration with large wingspan is primarily intended to lay a wide swath across long and broad fields, which does not require tight maneuvering. Ayer, Snow, and Weatherly also build low-wing monoplanes for the expanding agricultural air-

Schweizer Aircraft

4-3 *Schweizer Ag Cat Models G164A (left) and G164B.*

Piper Aircraft

4-4 *Piper PA-36 Pawnee Brave.*

craft market, and many operators keep overhauling their old Stearman trainer bi-planes, which were the original post-World War II Ag conversions. So both monoplanes and biplanes are used, depending on operational requirements.

Whether old or new, one wing or two, all agricultural aircraft fuselage structures have four things in common:

1 The payload hopper is located ahead of the pilot so that it absorbs energy without crushing him or her in a crash.
2 The structure is of welded tubing to protect against crashes.
3 Exterior panels readily removable for maintenance and wash down of spray and dust chemicals (Fig. 4-5).
4 All tubular structures, as well as control system components, are thoroughly corrosion-proofed for protection from the chemicals being applied.

4-5 *Schweizer Aircraft Ag Cat Model G164B ready for inspection or wash down.*

In fact, the fuselage structures are designed and built so ruggedly that there are known cases of midair collisions from which both pilots were able to transfer to different airplanes and continue spraying, having completely demolished the first pair of aircraft.

Landing gear components must be equally rugged to withstand overload operation from rough fields. Since it does not take long to load and spread chemicals, the Ag plane operates on a short cycle, averaging more landings per hour than most aircraft with the possible exception of flight trainers. Conventional landing gear with the main wheels forward and a tail wheel is preferred for rough field work, and is acceptable for professionally flown large aircraft.

The major source of Ag plane engines has been the World War II surplus market, with radials preferred because they are generally considered more throttle responsive than flat, horizontally opposed types. However, the required horsepower Pratt and Whitney, Wright Aero, Continental, and Jacobs radials are becoming scarce and therefore expensive. As a result other engine types are now being placed into production. The Piper PA-36 Pawnee Brave of Fig. 4-4 uses the 400-hp Lycoming flat eight, while higher-powered turbo-prop engines such as the PT6A-34 are now flown on some of the latest Ayers, Snow, and Schweizer models.

Cruising speed is not important because the Ag plane is not intended for cross-country travel and 90 mph is an acceptable speed for spraying or dusting. In similar manner climb is only important in regard to getting the maximum load off the ground, and altitude operation is not normally a design consideration. To carry, or rather, to stagger off with the huge payload of chemicals or fertilizers to be applied requires much horsepower and wing area, particularly on a hot, Gulf Coast summer day. So load-lifting capability is more important than low drag. Of course, dust and spray systems are an integral part and purpose of every Ag plane and differ slightly for each type in order to provide the optimum dispersal pattern. However, I'll leave the subject of chemical distribution to specialists in that field.

The important criteria for agricultural aircraft designed to work in tight areas are:

- slight positive stability plus high rate of roll for maneuverability;
- payload capability;
- operation from rough fields;
- a rugged airframe to reduce service and maintenance;
- accessibility for inspection and cleaning; and
- corrosion protection.

These requirements are satisfied respectively by the following:

- a short tail arm and the biplane configuration of short span plus four ailerons;
- wing area and plenty of power to carry the payload;
- rugged, conventional landing gear;
- welded steel tube fuselage;
- removable fuselage panels and wing inspection openings; and
- thoroughly primed and painted components, especially those subject to chemical spray or dust impingement.

Requirements are similar for the monoplane Ag plane designed to work the more open areas, except that maneuverability is not as important. As a result, the tail arm is longer, with adequate roll provided by a single set of large ailerons. You can clearly see how a variation in operational criteria has resulted in a configuration change among similar types of aircraft. The monoplane design was selected when operational requirements permitted because it is less expensive to build than a biplane.

Acrobatic and sport aircraft

The original pioneering spirit so essential to the early growth of aviation remains alive today in the sport-flying field. Pilots who enjoy flying for pleasure and re-laxation are turning in increasing numbers to homebuilt and custom aircraft of every configuration, with the open cockpit acrobatic design rapidly becoming a status symbol for this group. And if you've ever flown cross-country in an open cockpit during late spring, you know why this type of unrestrained flight is so appealing and satisfying. However, just what criteria do we apply to define a sport airplane?

There are two categories of aircraft in this market: the highly maneuverable acrobatic type and the stable, easily handled sports pilot's airplane. Although each satisfies large segments of the sport-flying market, their design and perfor-mance requirements are at opposite ends of the scale.

Acrobatic airplane design requires neutral longitudinal and directional sta-bility for high maneuverability, so a short tail arm is expected. Because the bi-plane offers wing area plus a high rate of roll for reasons discussed under Ag aircraft, this configuration is normally used; while well-balanced, responsive control surfaces are essential for smooth maneuvers.

To climb rapidly, a light-weight, powerful, and ample wing area is required. With low airframe weight and high power, balancing any airplane becomes a problem because of the engine weight, so it is not surprising that acrobatic air-craft have their engines and their pilot jammed back as close to the center of gravity as possible. As high speed is not essential, the drag of biplane wires and struts can be tolerated. In fact, it becomes an asset when making rapid transi-tional descents because the airplane comes down fast without picking up too much speed. To handle complex maneuvers that can load the entire structure, a limit load factor of 6 is used to design the airframe.

For climbing with much power at low speed, a large-diameter low-pitch or constant-speed prop is required, making acrobatic aircraft usually seem long legged. Because changes in flow direction felt by the pilot can be rapidly trans-lated into maneuver corrections, and for the pure sport of it all, the open cock-pit arrangement is preferred. To save weight, and again for the sport of it, the landing gear is normally a conventional type with a tail wheel. The Pitts Special of Fig. 4-6 is an outstanding example of the modern acrobatic biplane, and one which has received many first-place aerobatic awards in the United States and Europe.

Of course, acrobatics can be taught in aircraft intended for other purposes, provided the structure has been properly designed for such use or the maximum weight has been sufficiently reduced to permit higher load factors. For example, an airplane designed to 3.8 g loading at 2300 lbs (for normal category use) could be loaded to 3.8 g × 2300/6 g = 1457 lbs at 6 g. Although this weight might per-mit carrying little more than a student pilot plus instructor, two parachutes, and

Pitts Aviation Enterprises

4-6 *Pitts special acrobatic biplane.*

partial fuel, limited or full acrobatics would be possible provided the wing structure is capable of taking sufficient negative (down) loading.

Acrobatics should never be performed in a normal category airplane unless the flight manual permits such operation at reduced weight. If the structure has not been suitably reinforced, some unexpected and spectacular acrobatics will result if the wing fails under down loading.

A few standard production aircraft have been redesigned and strengthened for student acrobatic instruction, notably the Cessna 150 Acrobat and the Bellanca Citabria. Also, the Pitts biplane has been expanded to permit tandem seating for acrobatic instruction and pleasure flying. These three models are the current standard acrobatic training aircraft.

A new series of acrobatic aircraft is now emerging in a monoplane configuration. Although this type has been more popular in Eastern Europe than in the United States, development models of acrobatic monoplanes are being built and flown here with some success. In fact, this special monoplane is now becoming the preferred vehicle for our most advanced acrobatic pilots.

Based on competition results, most successful acrobatic aircraft have short span biplane wings for maximum area and rate of roll, light structural weight, and a short coupled fuselage mounting a large engine close to the center of gravity. Because of the large-diameter propeller required to absorb power and provide a high rate of climb at low speed, the conventional landing gear tends to pitch the nose up at rest, which is also the three-point stall landing attitude. Include coordinated and responsive controls, an open cockpit for wind-flow sensing and pleasure, and we have the design formula for practically every acrobatic biplane design.

This concept has also been carried over into the pure sports pilot's airplane, as indicated by the great amount of homebuilt biplane activity among Experimental Aircraft Association (EAA) members. Totaling more 300,000 active aircraft enthusiasts, the EAA has sponsored the design and sale of plans for constructing the easily handled ACRO Sport biplane and the Pober Pixie monoplane designed by former EAA President Paul Poberezny and shown in Fig. 4-7. A combination of high costs for factory-built airplanes, creative interest, and love of flying, has resulted in literally thousands of homebuilt aircraft being built at this time. Many designers offer plans and kits for a variety of biplane and monoplane aircraft of conventional and unconventional configuration.

Sport Aviation Association, photo by Lee Fray.

4-7 *Experimental Aircraft Association Pober Pixie sportplane.*

To be safe and successful, the sports pilot's homebuilt airplane differs from the purely acrobatic type in that it should be easy to fly. Many homebuilders do not have any pilot rating and expect to learn to fly in the airplane they are carefully constructing; while others are only occasional pilots looking forward to the satisfaction of flying the airplane they are building. One of the best engineered and most efficient homebuilt landplane designs is Ladislao Pazmany's PL-2 shown in Fig. 4-8. This airplane has also been built in quantity as a trainer for the air forces of Taiwan, South Vietnam, South Korea, Indonesia, and Ceylon.

Pazmany Aircraft

4-8 *Pazmany PL-2 homebuilt design.*

While this design is of all-metal construction, other types available are fabricated from wood, steel, or aluminum tubing with fabric or metal covering, plastic foam with fiberglass skin, and combinations of all methods. Through such experimental structural development the EAA membership is making a direct contribution to the private aircraft industry, since some acceptable form of composite structure will eventually be evolved to reduce airframe construction labor-hours and cost. At the present time, FAA design criteria unfortunately tends to penalize the use of composite plastic construction for certified production aircraft, but this situation will change as acceptable methods of composite construction are developed.

An increasingly large and interesting part of sport flying centers around the reconstruction or duplication of "golden oldies"—antique aircraft produced during the late 1920s or early 1930s. While an unbelievable amount of time, energy, and money can be spent tracking down and rebuilding old components, the final results can be worth the effort as shown by Ed Wagner's Grand Champion Antique Award-winning American Eagle of Fig. 4-9.

Similar efforts have revived original Wacos, Great Lakes Trainers, DeHavilland Moths, Fairchild, Stinson, Taylorcraft, Piper, and Champion aircraft because all these planes are stable and relatively easy to fly. And that is the prime consideration for the sports pilot's airplane. This means adequate stability, gentle stall, low landing speed for short-field performance, good angle of climb, moderate cruise, and responsive controls. To provide these characteristics, the acceptable sport airplane has:

- a fairly low-wing loading of 6 to 12 lb/sq ft;
- a power loading of 10 to 22 lb/hp;
- combined wing and power loading totaling less than 30, and preferably less than 25;
- long tail arm;
- positive longitudinal and directional stability with forces about equal for all control surfaces; and
- responsive controls.

Useful load capability and cruise performance must be adequate, but are not of prime importance because this type of airplane is used principally for local pleasure flying or short cross-country hops.

4-9 *American Eagle antique airplane.*

Sailplanes

Although this book is primarily concerned with airplanes, which by FAA definition are engine-driven fixed wing vehicles, I've included some review of sailplane design because motorless aircraft fascinate everyone—pilots and non-flying public alike.[1]

Unpowered aircraft were originally called gliders because all they could do was glide downhill. Due to a combination of high drag, inefficient wing sections, relatively heavy wing loadings, and poor pilot technique, the first gliders had very limited performance.

Following World War I, German aeronautical engineers were restricted to motorless flight. In order to develop pilots for Goering's Luftwaffe, German technical effort was concentrated on the development and construction of great numbers of increasingly sophisticated glider designs. Eventually the state-of-the-art reached a level permitting sustained flight by use of ascending air currents—and the sailplane was born. Today's sailplanes fall into three categories: trainer, intermediate, and high-performance or competition designs. Under proper conditions any of these types is capable of soaring flight; they are principally defined by glide ratio and the degree of pilot skill required for safe operation.

The Schweizer 2-33A of Fig. 4-10 is the standard U.S. two-place trainer and demonstration ride sailplane. With steel tube, fabric-covered fuselage and tail surfaces plus an all-metal wing, the 2-33A is sufficiently light to soar readily with one aboard, and has taken two people more than 10,000 ft above the launch point in strong lenticular wave lift conditions. This sailplane has a glide ratio approaching 20 to 1 at 45 mph and is relatively easy to fly. The stall speed of about 35 mph indicates that things happen slowly and with minimum impact energy. As a result, a pilot has to be quite lax to get into serious difficulty with this trainer.

Schweizer Aircraft

4-10 *Schweizer 2-33A training sailplane.*

The intermediate class sailplane is usually a single-place design offering a glide ratio between 25 to 1 and 30 to 1. This type is capable of limited cross-country flight as well as ridge and thermal soaring, and might be equipped with oxygen for altitude work. The Schweizer 1-26 sailplane is one plane in this category and is the largest single class of active one-design sailplanes in the United

States. In fact, the Schweizer 1-26 regional meets and national contests are held in very similar fashion to one-design yacht competitions. The intermediate sailplane is somewhat more responsive and sensitive than a trainer and requires a higher level of in-flight planning.

The high-performance design is usually a single-place type, like the one shown in Fig. 4-11, and has a glide ratio between 30 to 1 and 40 to 1. Along with this high glide ratio, things happen very fast and elevator control response is quite rapid and sensitive. In order to obtain minimum drag, the tail load has been kept very low, resulting in essentially neutral or slightly positive longitudinal stability. As mentioned in chapter 2, at a 40 to 1 glide ratio, an 800-lb sailplane has only 20 lbs of drag. So it is slippery and capable of rapidly picking up speed, which can be quickly converted into high g loadings in the hands of a novice sailplane pilot.

Schweizer Aircraft

4-11 *Schweizer 1-35 high-performance sailplane.*

Regardless of the design category, certain features are common to all sailplanes. Since it is necessary to remain in a rising thermal as long as possible to gain altitude, and as most thermals are of small diameter, the ability to circle tightly and slowly with minimum drag is most important. Higher-performance designs achieve this by use of high aspect ratio wings, long tail arms, and neutral stability. The resulting light elevator forces are also of benefit as they are less tiring during long duration flights. However, neutral stability flight requires a relatively high degree of pilot skill, so increased longitudinal stability with its associated higher drag is expected in trainer and intermediate designs compared with competition sailplanes.

High aspect ratio is also an important factor in obtaining high lift-to-drag ratio flight, because drag caused by wing lift is reduced in direct proportion to increased aspect ratio [which is the ratio of wingspan squared divided by wing area]. This, of course, explains why sailplanes have large span, narrow chord wings; the designer is reducing drag by increasing wingspan for a given area. This subject is thoroughly covered in chapter 6, and so will not receive further comment at this time.

Due to long span wings, adverse yaw (the tendency for aileron drag to turn the ship opposite to the desired direction) becomes an almost incurable problem. This condition is overcome by leading with the rudder and applying rudder as necessary during turning flight. To obtain maximum response at minimum drag penalty, a long tail arm and effective rudder are necessary. It would also seem that some additional design effort could be directed toward either reducing aileron adverse yaw or developing a more effective lateral-control system.

Because of relatively low sailplane drag and the accompanying ability to exceed placard maximum speed, some form of speed control is desirable. Such control has typically been included in the form of wing-mounted dive brakes capable of limiting speed in a dive condition. Adequate speed control has most recently been provided by plain trailing edge flaps that can be lowered slightly for thermalling, raised a bit to reduce wing section moment and drag during cruise, and dropped up to 90° for landing approach control.

Competition sailplanes also feature a retractable landing wheel, faired cockpit canopy, and water ballast to increase wing loading and cruising speed for a given L/D (lift-to-drag) value. Since these are cross-country aircraft, they carry some radio and duplicate sensitive rate of climb (variometer) equipment.

Sailplanes have been developed to a point where all designs now look very much alike. The successful high-performance single-place sailplane typically has an aspect ratio of more than 20 to reduce wing drag, a tail arm of around five times the average wing chord, tail surfaces having a smaller percentage of wing area than is normal for powered aircraft due to the longer tail arm (basically reducing surface drag), neutral to slightly positive longitudinal stability at an aft center of gravity to reduce tail load and drag, responsive ailerons with a high rate of roll to permit thermalling flight, and some form of speed and approach (sink rate) control surfaces—either dive brakes or flaps. Of course, the fuselage and all intersections must be of excellent aerodynamic design for competition types, although this is a bit less crucial for intermediate designs and much less so for trainers. With such rigid criteria it is no wonder that sailplanes look alike.

Seaplanes

I could go on for hours reviewing seaplane design criteria because amphibians have been my particular interest for 50 years, but seaplanes are covered in-depth in chapter 15. To provide a background for those of you not seriously interested in the design of watercraft, the three basic types of seaplanes are discussed here with a minimum of detail. These three basic watercraft are the floatplane, the hull-type flying boat, and the amphibian, which can be of either float or hull configuration equipped with landing gear.

The Cessna 172 Skyhawk of Fig. 4-12 is a typical example of an airplane that was originally designed for land operation converted into a seaplane. This conversion capability explains much of the popularity of seaplanes; for relatively small additional cost an owner can enjoy water operation during the summer, returning to wheels or skis when freezing weather sets in. Many seaplane pilots consider the floatplane easier to dock than hull-type flying boats, which adds to the acceptance of the floatplane. However, this is not necessarily so, as you will see in chapter 15.

4-12 *Cessna 172 Skyhawk with EDO 89-2000 floats.*

A hull flying boat looks like the Teal amphibian of Fig. 4-13 but with the landing gear removed. I could not include a picture of a modern flying boat because there has been no pure hull seaplane produced since the late 1920s except for the large ocean-crossing clipper types of the 1930s and 1940s. The small boat went out of production primarily because the cost of building a seaplane hull could not be justified for limited water operation.

Cost considerations were magnified by the belief that landplanes could be mounted on floats to do the same job as seaplanes. However, converted landplanes cannot do the same job as a hull seaplane when operating in rough sea or severe wind conditions. This is because the floatplane is poised high above the water with no outboard wing float protection and, therefore, is much more vulnerable to capsizing than is the hull-type seaplane.

The hull seaplane survives today in the amphibian configuration capable of operating off land or water. This type of utility airplane was popular before and just after World War II in various twin-engine designs developed by Grumman, and remains in production today in single-engine versions such as the Skimmer, Buccaneer, and Teal amphibians produced by different small companies.

The amphibian is a design concept that offers a unique degree of utility from both land and water operation. The amphibian, as shown in Figs. 4-13, 4-14, and 4-15, has a hull plus retractable wheels, or wheels in floats for water operation. Because the hull chines and transverse step are required for water work, amphibian drag is greater than that of a landplane of similar gross weight; and so the rate of climb and cruising speed are lower. Since water contact at 60 mph has an impact effect like that of a concrete surface, the hull bottom skin and frames must be much heavier than for a landplane structure, adding to hull weight and cost.

Howard Levy

4-13 *Teal amphibian with the first T-tail approved by FAA for Part 23 aircraft.*

Lake Aircraft

4-14 *The Lake Buccaneer amphibian.*

Floats with retractable wheels as shown in Fig. 4-15 are capable of providing amphibian operation through a fairly simple exchange of landing gear. However, amphibian floats are heavy and costly, generally resulting in a conversion that is more expensive than an amphibian flying boat of similar capacity. Also, it is much more difficult to handle in land crosswind landings or severe wind conditions on water because it sits so high off the surface.

Whether of float or hull type, seaplanes must be designed to function well on water because that is the primary reason for their existence. This means that modern seaplanes need the following characteristics:

- very mild or no porpoising limit characteristics (that is, no severe oscillation in pitch at high or low hull trim angles on the water during takeoff or landing);
- no tendency to skip (pitch out of the water during landing);
- The ability to perform step turns of small radius in order to operate in small bodies of water;
- be controllable on the water for precise sailing or steering when docking; and
- be able to take off with a minimum water run on a hot day.

Adequate control at low speed is also essential during the unstable transition from a displacement boat to a planing seaplane. Since this takeoff phase, known

EDO Commercial Corporation

4-15 *Cessna 185 Skywagon with EDO 597-2790 amphibian floats.*

as "getting onto the step," occurs around 20 mph, ailerons and elevator surfaces must be fairly large to be effective at this low speed.

All of this, naturally, costs in some fashion. Water performance requires power and wing area—there are no substitutes—and both are heavy and expensive. A watertight and rugged hull is also heavy and expensive, but landing on water at 60 mph is quite similar to landing on a runway with the gear retracted. So the hull must be strong. For operation around saltwater, a corrosion-proofed structure and control system are essential, although more costly than similar landplane components. And the large movable surfaces required for low speed control might feel heavy at cruising speed.

Seaplanes are used principally by private or sports pilots and bush operators. This means a stable design is required, generally following the criteria set down for private aircraft. In addition, whether a hull flying boat or floatplane design, the seaplane is more expensive and carries less useful load than a landplane of equivalent weight. To be effective, it should also have more horsepower. In return, the seaplane offers increased utility plus the joy and freedom of water flying. Possibly most important of all, the boat-type amphibian offers the combined safety of multiple landing fields and the crash protection of a rugged hull.

Despite cost, cruising speed, and reduced useful load limitations, small amphibian performance has developed to a degree permitting limited volume production with three designs currently certified for FAA requirements. After many years of frequently discouraging effort, amphibian utility is slowly but steadily being recognized as sales increase from year to year.

Just as riverboats opened nineteenth-century frontiers, the amphibian will soon be used to expand commerce in the developing countries of Africa, Asia, and South America where lakes and inland waterways are plentiful but airports are not.

Operational requirements

As this review should clearly indicate, a number of different aircraft designs are required to fill various operational missions. But whether aircraft are designed for private flying or specialized use, they must satisfy certain basic flight parameters if they are to receive pilot acceptance.

While the order of importance or weighting of items can vary with owner preference or actual use, the primary criteria used by the average general aviation pilot to evaluate powered aircraft are:

1 crosswind handling
2 rate of climb
3 adequate stability for IFR operation and approach procedures
4 useful load
5 cruise performance

For some applications, cruise performance can be more important than useful load, in which case criteria 4 and 5 would be reversed. But, regardless, the first three items directly affect flight safety, and they are more crucial than the last two.

Adequate crosswind handling not only designs the control surfaces, as you will see in the next chapter, but also directly affects flight safety near the ground—the most crucial flight regime. In similar fashion, rate of climb has a direct bearing on the level of flight safety offered by any airplane. This is evidenced, for example, by the ability or inability to climb out of short fields or over hills bordering an airport or lake; ceiling capability in mountainous or icing conditions; missed approach climb response; and hot day operation. Criteria 1 and 2 obviously have a major effect on safety.

Criteria 1 and 3 combine to provide an easily handled airplane, which also directly relates to safe operation. Responsive controls and inherent stability about all three axes are essential for IFR work. Fortunately, meeting criterion 3 also results in an airplane that is pleasant to fly in stable air and one that does not become a demon in turbulent conditions.

Useful load is always marginal, so the degree of inadequacy is the determining factor for criterion 4. However, it would help if aircraft advertised to carry four adults could carry four 170-lb people plus 100 lbs of baggage in addition to full fuel, oil, primary blind flight instruments, and basic IFR radio equipment. This would require a total useful load of about 1130 lbs.

Currently, no production 150-hp four-place airplane can carry the target weight. The Cessna 172 Skyhawk and the Piper PA28-151 Warrior come closest and, not surprisingly, enjoy the largest production volume of the fixed gear, 150-hp class. However, at gross weight on a hot day their climb performance leaves something to be desired. The 180-hp versions of each improve this important requirement, while also slightly increasing useful load and cruise, as well as cost.

On the surface, high-speed cruise performance for private aircraft seems crucial only if frequent trips are to be made between distant locations of perhaps more than 300 statute mi. After all, the ability to fly at 100 mph in a straight line usually more than doubles automobile cruising speed, because major highways seldom connect desired destinations by direct routes. Of course, winds aloft do affect ground speed and always seem to be blowing from the desired heading. In view of this, the higher the economical cruising speed the better. Because no one enjoys listening to a 2500-rpm power setting for hours on end, the ability to cruise small aircraft at 130 mph or faster at 2300 rpm or less can be most beneficial from both the sound level and fuel economy standpoints.

There is much merit to having a relatively high cruising speed at fairly low power. A brief ground speed example illustrates this quite clearly: With a frequently experienced 30-mph headwind condition, a 110-mph cruising speed delivers only an 80-mph ground speed; however, a 150-mph cruise speed results in a 120-mph ground track—a 50% increase in distance covered for a 36% increase in cruising speed. So design cleanliness is important, even for small private aircraft, and particularly so when fuel prices seem to increase faster than performance.

Now, how does the private aircraft designer achieve all these goals? Some are in direct opposition. For example, high cruising speed vs. low approach and landing speeds. High cruise usually means low wing area as well as plenty of power, while low stalling speed requires much wing area and/or low gross weight. But with high power, we have an expensive engine and a heavier one, adding to cost, weight, and landing speed.

If the airframe is too heavy, the useful load is too low to be practical. If too much wing area is sacrificed to obtain high cruising speed, both short-field capability and climb is compromised. If too much power is used to obtain climb and cruise for a given gross weight, the useful load goes down as fast as the cost goes up. If too much wing area is added to obtain short-field capability, the cruising speed decreases, while the weight and cost increase.

In view of the preceding, it is not surprising that designers study similar successful aircraft before starting on any new project. Compromise is obviously required, with trade-off decisions determining the level of acceptance for the resulting airplane. To define and describe aircraft design procedures, we must be concerned with just how such decisions are reached, along with the effect of various options on aircraft performance and handling characteristics.

So far, these introductory chapters have provided background information relating to all general aviation aircraft. Next, I'll consider the practical application of design requirements for private aircraft beginning with particular emphasis on criteria 1 through 5 of this chapter, which is presented in that order, with two chapters devoted to criterion 3. To be more precise, I will review the effect of various design criteria on the handling and performance characteristics required by small aircraft. I'll also consider why these aircraft must look as they do to satisfy FAA regulations—and, more important, to be accepted by private pilots. For it is pilots who determine whether or not a new airplane will be successful. In the final analysis, it is the customer who designs the airplane.

Endnote

[1]FAA FAR Part 1: Definitions and Abbreviations, p. 1.

5

Handling crosswinds

Design requirements

Crosswind conditions usually affect the beginning and end of every flight. Although the landing phase is generally more critical, things can get quite active during takeoff if the rudder and ailerons are not effective at low speed. This situation can be bad on land, and can be dangerous on water, so responsive controls are essential for satisfactory crosswind operation.

When near the runway surface, it is most reassuring to have excess aileron control available to handle crosswind gusts and changes in wind direction. At low speeds, ample aileron power during flare can correct momentary trim changes in roll and yaw, while marginal or insufficient roll control can result in premature ground contact plus accompanying bounce, ground loop, or wing-tip damage.

If the ailerons are too small in chord or span, when deflected downward they cannot generate enough increase in lift coefficient (through change in wing-section camber) to provide the lift required to raise a low wing. Conversely, when deflected upward small ailerons cannot reduce lift to the degree necessary to rapidly lower a high wing. Since the ailerons are interconnected and act together to provide lateral control, it is obvious why weak ailerons cause a marginal rate of roll and could make an airplane impossible to handle in gusty crosswind conditions.

To be effective in changing wing camber, an aileron needs a chord approximately 15 to 20% of the wing chord aft of the hinge line, and a total area (both ailerons) of about 7 to 10% of the total wing area as shown in Table 5-1. Of course, too much area makes the ailerons feel "heavy" to operate. In addition, the outer end of the aileron should be at or just inboard of the wing tip. This location places the center of aileron area and lift as far from the airplane centerline as possible to provide maximum rolling moment about the longitudinal axis. Note that increasing wing area by adding span outboard of the ailerons effectively buries them, resulting in an undesirable reduction in roll rate and control response.

Table 5-1   Aileron area and power

(1)	(2)	(3)	(4)	(5)	(6)
	Wing area S_w	Aileron area $S_{A_{total}}$	$\dfrac{S_{A_{total}}}{S_w}$	℄ Airplane to center of aileron area, ft	Aileron power
Aircraft	Sq ft	Sq ft			3 × 5
SIAI Marchetti SF 260 (260 hp)	108.5	8.50	.078	11	93.5
Czechoslovakia Zlin Z-526L (200 hp)	166.3	14.8	.089	13.2	195.4
Wassmer Super 4/21 (235 hp)	172	13.35	.078	13	173.5
Jurca Sirocco MJ-5H2 (160 hp)	108	10.0	.093	9	90.0
Cessna Cardinal (180 hp)	174	18.9	.108	12.9	243.8
Piper PA28-151 Warrior (150 hp)	170	16.1	.095	12.4	199.6
Piper PA28-200R (200 hp)	170	12.4	.073	11.5	142.8
Robin HR 100 (180 hp)	176.5	13.9	.079	13	180.7
Thurston Teal (150 hp)	157	12.8	.081	13.3	170.2
Sequoia S285 (285 hp)	130	13.8	.106	11.9	164.2
			.088	Average	165.4

The wing panel can be tapered from root to tip to further increase roll capability, or possibly only in the outer portion of the span ahead of the ailerons. Reducing the outboard area compared with a rectangular wing decreases wing panel damping action, and so makes the ailerons more powerful for a given aileron area and spanwise location. For example, we all know that a small paddle can be moved through the air more easily than a large one.

To improve approach and landing control, adverse yaw needs to be kept to a minimum. Adverse yaw is the tendency of the airplane to turn to the right when left aileron is applied, and to the left when using right aileron. Adverse yaw is caused by the increased camber drag of the down aileron and reduced camber drag of the up aileron. It is usually most pronounced and difficult to manage at relatively low airspeeds occurring during corrective maneuvers in the landing flare. Adverse yaw is particularly magnified at lower speeds because fin and rudder effectiveness decrease rapidly with airspeed; in a falling-dominoes-type reaction, reduced fin effectiveness causes a loss of directional stability and, therefore, the ability to hold a desired heading by correcting for adverse yaw.

Because decreasing control response is obviously undesirable at this critical point in the landing procedure, control effectiveness must be improved to the extent possible through modified aileron and vertical tail design. The Frise aileron and associated variations of this type have a section forward of the hinge line which projects down into the airstream below the wing when the aileron is raised. This action has the aerodynamic effect of dragging back the wing panel having an up aileron, thereby turning the airplane into the direction of bank called for by aileron deflection. Most aircraft have adaptations of this type of aileron design, as can be noted during your next preflight inspection.[1]

No discussion of roll control is complete without noting the effect of dihedral—a full subject in itself. Dihedral is the spanwise angle the wing chordal plane makes with the horizontal, and is usually greater for low-wing aircraft than for high-wing designs, as shown in Fig. 5-1. A larger dihedral angle is necessary for low-wing aircraft to bring the center of lift closer to or above the center of gravity. This location increases static stability about all three axes by having the lift force act above the airplane weight centered at the center of gravity. This effect is clearly shown in Fig. 5-2, as well as by Figs. 2-4 and 3-4. The center of gravity is below the lift center of high-wing aircraft, providing a form of built-in pendulum stability which assists in righting the airplane in roll and pitch. This corrective action is a measure of static stability, which is the ability of an airplane to return to trim attitude when displaced from level flight. The faster and more complete the return, the more stable the airplane. Usually, the further the center of gravity is located below the center of lift the more stable an airplane is in cruising flight. This is a major and distinct advantage of high-wing aircraft.

5-1 *Dihedral angle.*

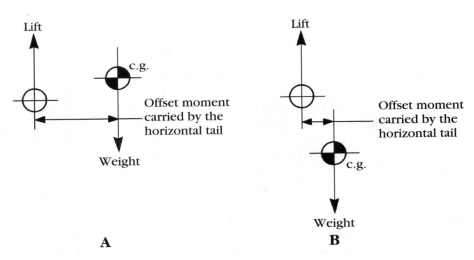

5-2 *The effect of lift and weight forces on static stability. (a) Low wing position. A relatively unstable condition. The corrective action of the horizontal tail prevents the lift and weight forces from rotating over 90 degrees to become vertically oriented with the center of gravity directly below the center of lift. (b) High wing position. A relatively stable relationship requiring very little horizontal tail load to balance the few degrees of angular rotation that would align the lift and weight forces. This force relationship provides pendulum stability as explained in the text.*

While dihedral is related to both the lateral and directional stability characteristics of the airplane, too much dihedral will make precise control of the degree of roll most difficult in gusty conditions. This is so because the down (windward) wing has more projected area and provides more lift about the airplane centerline than does the raised (downwind) wing. As a result, more aileron power is needed to keep the windward wing down on any design having too much dihedral. For crosswind control, moderate dihedral is required for stability, with somewhat more aileron power necessary for low-wing designs.

To satisfy these requirements, we find our most responsive crosswind aircraft fitted with generous but not overly heavy-feeling Frise-type ailerons positioned as far out on each tapered wing panel as practical. Not surprisingly, this configuration is found on all of today's most popular private aircraft.

The level of directional stability provided by the fin and the amount of rudder control available both decreases with airspeed. In fact, if approaches are made with plenty of flap down and decreasing speed, rudder control is usually in short supply sometime during the flare, and probably before the airplane runs out of roll control. This can be due to partial blanketing of the vertical tail by the wing and flaps, or possibly from turbulent flow off the cabin structure. Fortunately, the slipstream effect while applying power usually restores directional control quite rapidly.

The use of a T-tail configuration not only makes the fin and rudder more effective due to the end plate action of the stabilizer location, but also positions the horizontal tail above wing turbulence. The addition of a dorsal fin tends to straighten turbulent flow over the fin and rudder, increasing the effectiveness of both while providing additional fixed fin area to improve directional stability at low speed. The new Beechcraft model 76 twin of Fig. 5-3 incorporates both of these features.

5-3 *Beechcraft Model 76 light twin with T-tail.*

The so-called uncontrollable deep stall experienced by heavy transport and military aircraft has been frequently blamed on the T-tail configuration. Actually high-wing loadings and powerful flaps are the real cause due to the resulting strong downwash flow off the wing and over the high stabilizer. Deep stall should not be experienced by light aircraft having a T-tail design because relatively low-wing loadings cannot establish a downwash flow of sufficient energy to overcome pilot effort, and thereby lock the elevator in the up position. A few designs are now flying with a dorsal fin and T-tail combination, although this horizontal tail position is really a special purpose location. Information on tail surface design is found in chapter 7.

The crosswind landing gear, further described in chapter 14, is an interesting post-World War II effort to solve some of the more serious crosswind problems that developed as grass airfields were replaced by asphalt runways. A number of different systems were patented, varying principally in caster design. This improvement was intended for conventional gear aircraft having the main wheels forward plus a tail wheel. The rudder steered the tail wheel, and was usually designed to swivel free after turning 30° to the right or left.

Crosswind landing gear, or *self-castering gear*, was thoroughly demonstrated during the mid to late 1940s and flown by many pilots at all levels of proficiency. This unique main gear design permitted aircraft to land with a crabbed heading, and then run out in a skewed fashion with the airplane pointed one way and the wheels another, as shown in Fig. 14-3. This was accomplished automatically at ground contact by having the main wheel axles "break free" under a side loading of predetermined value.

Unfortunately, the gear produced some odd effects—slewing off the runway if a turn occurred at a speed high enough to permit castering. A few of these negative handling characteristics, plus rapid acceptance of tricycle landing gear, terminated further development of crosswind gear during the late 1940s.

One very promising approach to solving the flight control coordination problem, the two-control airplane, was found deficient and dropped, primarily because of crosswind operational requirements. Shortly after World War II, the lightplane industry was vigorously pursuing this so-called two-control airplane.

A two-control airplane has interconnected ailerons and rudder plus normal elevator control. This type is best remembered for the still active Ercoupe designed by Fred Weick just before the war and produced in great numbers after 1946. However, as time passed, the crosswind shortcomings of two-control aircraft in the hands of the average pilot became all too obvious. It was not possible to bank into a stiff crosswind to check drift without also flying off the runway heading, and any crab angle could not be straightened at touchdown in an airplane without independent rudder control. Fred Weick says his solution was to bank into the desired heading, level the wings, and land at the flight crab angle. He never had a landing gear failure in his Ercoupe; so this can be an acceptable crosswind technique for all two-control aircraft, provided the landing struts can take the side loading.

Although some two-control aircraft were a delight to fly, producing ball-centered turns with an automotive-type steering wheel (so anyone who could drive could also expect to fly with minimum practice—or so the ads declared), two-control aircraft became a real handful for many during approaches in a moderate crosswind. As a result, serious attention returned to the three-control system of ailerons, elevator, and rudder which is now universal for private aircraft.

Having reviewed the effect of control surfaces on handling response and flight characteristics, let us now run through a set of crosswind operations to see just what happens when we apply these controls in flight.

Crosswind flying

Many of us were first attracted to flying when exposed to the challenge of mastering flight, possibly as a passenger in a friend's airplane or during the thrill of an introductory ride. And there is no better way to keep this mastery current than by practicing crosswind takeoffs, approaches, and landings.

All too many pilots, particularly those who only fly a few hours each month, retreat to "hangar flying" as soon as the surface wind gets much greater than 15 mph or approaches an angle of 40° to the runway. Flying is not a particularly difficult art but it does require practice and acquired skill, especially in adverse or less than ideal conditions.

We tend to shy away from crosswind flying for many reasons. If the winds are high, the air is likely to be gusty and turbulent due to frontal passage and the wind angle varies all over the field along with the gust velocity. Certainly not a relaxing situation and not one easy to sit out, particularly when the tower reports winds leading into the danger zone as shown in Fig. 5-4.

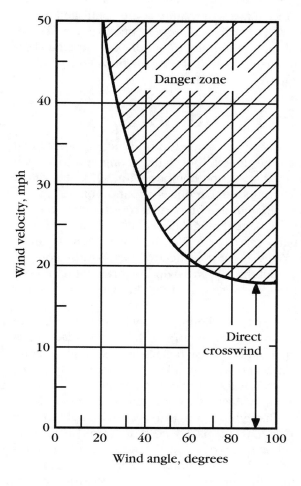

5-4
Maximum safe crosswind velocities.

We all know that a 30-mph wind at 60° to our runway requires correction on downwind and base but we are quite far from the ground then and feel relatively safe. The conditions seem to magnify as the runway approaches (or drifts to one side) during final. But the time to make corrections was during the early phase of the downwind leg when the pilot needs to track (fly over the ground) parallel to the runway, and so be prepared for the base and final legs.

Some pilots have been advised not to use flaps during fairly high or gusty wind conditions. Flaps can and should be used on all approaches because they act to stabilize most aircraft in both pitch and roll, as well as reduce ground contact speed. However, the degree of flap application varies with the airplane type and with surface conditions, and so presents another decision for many pilots.

Ground handling during taxi and takeoff also becomes an increasingly demanding problem in high winds, and in fact can require some controlled rate of reducing aileron displacement. No test of skill, but frequent practice would help. Since most pilots are concerned about their ability to handle the airplane competently, let us take a further and closer look at the much magnified and frequently maligned crosswind operational requirements.

Crosswind conditions

As shown in Fig. 5-5, any crosswind can be divided into two components that act on the airplane: one counter to the direction of flight, acting at a lower velocity than the surface wind; the other, a second component at 90° to the direction of airplane motion. The first component reduces ground contact speed, similar to any headwind, while the right angle component tries to make the airplane drift, weathercock, or flip over—either singularly or all at once.

For example, with a surface wind velocity of 30 mph blowing at a 60° angle to the runway heading (Point 1 of Fig. 5-5), the effective headwind (reducing ground speed) is 15 mph (Point 2), but the 90° crosswind component trying our skill is 26 mph (Point 3). Small wonder these conditions lie within the danger (alert) zone of Fig. 5-4.

We have a few tools to combat such conditions, and with practice we can operate in fairly severe weather as though on a normal flight, provided heavy crosswinds are never treated as a routine matter. In fact, it is strongly recommended that any initial practice in severe crosswind conditions be under the guidance of a competent instructor, preferably sitting in the right seat.

Although all controls participate fully in crosswind work, rudder plus aileron coordination really come into play first when setting up the crab angle to hold course on downwind, base, and final. There is no need to crank in upwind aileron 800 ft above the airport on final; why develop a shaking set of calf muscles from holding opposite rudder for 2 min or more to keep centered on the runway? Further, during all this time the airplane has crossed controls, so any sudden drop in gust velocity, a sharp upgust, reduction in approach speed due to inattention, or turbulence-induced trim change can produce spin conditions. Certainly a spectacularly unpleasant way to end any flight.

A much better practice is to keep the runway centered in your sights by maintaining a comfortable crab angled approach, power off or low, coupled

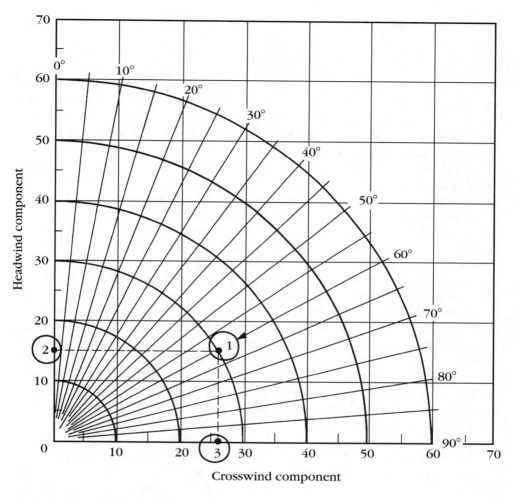

5-5 *Wind component chart.*

with a steady descent speed and rate controlled by the elevator. About 50 ft above the runway, aileron is applied to drop the upwind wing to stop drift while at the same time opposite rudder is used to keep the nose straight and the runway centerline sighted. Apply more or less rudder and aileron as necessary to check immediately any runway centerline drift downwind or upwind as gust conditions vary.

Commence to flare and follow through to land on the upwind main wheel, then the downwind main wheel, and finally the nosewheel or tail wheel. Before contacting with an interconnected, steerable nosewheel, it is advisable to get the rudder centered or you can go shooting off the runway downwind (due to holding opposite rudder which casters the nosewheel downwind). However, upwind aileron should be held during the runout.

An alternative crosswind approach procedure holds the upwind crab angle during flare, with a final downwind rudder kick used to straighten everything out just before touchdown. I was taught to fly my PT-19 this way, and it is great sport and truly a test of skill to execute this correction maneuver properly and at the precise moment required. But only if the wind is light and landings are made on a nice grassy surface. When operating in a high crosswind and landing on an asphalt or concrete runway, watch out!

If the crab is not fully corrected, the combination of upwind facing nosewheel plus ground speed could ground-loop a tricycle gear airplane as surely as a tail dragger. And if the crab correction is kicked out too high above the runway, downwind drift will thoroughly test the side load capability of the main gear struts. As a final insult, too much downwind rudder correction at touchdown can result in running away downwind. These maneuvers are described in more detail in chapter 14, Landing gear design.

For the average pilot, crosswind conditions are more easily handled with tricycle landing gear, preferably with the spanwise distance between wheels, known as the tread, approaching one quarter of the wingspan. The wider the tread, the greater the angle of bank possible to handle crosswinds before dragging a wing tip. While a high-wing design appears capable of steeper banks, and thus being able to operate in more severe crosswind velocities than low-wing types, this is not the case. Nor is it a critical consideration because low-wing designs normally have the main gear located well out on the wing. Either configuration is equally effective in crosswinds provided sufficient approach and flare speed is maintained to assure aerodynamic control.

When flying speed is too low for crosswind conditions, the airplane soon lets you know by running out of rudder and aileron control. When this happens, either increase speed to regain control, or apply power and go around again; running out of rudder or aileron can be an eerie and long remembered experience.

Recommended approach and primary flight control procedures have purposely been discussed separately from flap operation. Flaps should be used to some degree during final approach and possibly when on base leg, depending on height and speed in the landing pattern, wind strength, and runway length available.

Just how much flap to apply is a matter of judgment influenced by airplane handling characteristics as well as by surface wind velocity and direction. The use of 10 to 15° of flap to improve stability during the early portion of the final leg, and increasing to 20 to 30° during the final half would seem a good compromise in crosswinds blowing up to 20 mph and angled up to 60° with the runway. Possibly more flap could be used for shorter runways. A few practice landings will provide the answer for you. But remember to keep flying speed higher than usual with flap operation in order to maintain an adequate margin of control. Since the wind is blowing harder than normal, the headwind component reduces ground contact speed. Also, the extra flying speed is most helpful when a gust drops 10 to 15 mph or changes heading 30°. At any rate, full flap deflection is excessive under extreme crosswind conditions. Also, be sure to dump the flaps once finally on the ground; the low contact speed obtained by using flaps assures that any normal gust cannot lift a wing once the flaps have been retracted.

Floatplanes, flying boats, and amphibians have somewhat special problems during high wind conditions. When landing on the average waterway, any seaplane can usually select an approach path heading directly into the wind and execute a full stall landing at relatively low surface contact speed. However, when water approaches must be made in crosswind conditions due to a narrow canal or creek, the amount of crosswind that can be handled is somewhat limited by wing floats in the case of flying boats and amphibians, and equally so for float seaplanes by the prospect of tripping the upwind float on wave tops.

Hull-type amphibians are also limited by possible wing float contact with the runway surface when operating in crosswinds on land, but this is not particularly serious because the landing gear is (hopefully) extended and provides additional ground clearance.

As a final note, all crosswind approaches and landings are frequently made even more difficult by the presence of upwind trees, buildings, embankments, snowbanks, and other obstructions. These obstructions create both additional turbulence and rapidly variable wind headings. This same effect is noted to a lesser degree during final approach, when course correction can be required to accommodate variations in wind direction and velocity with altitude. Any IFR pilot knows this only too well from trying to keep the localizer needle centered during the final moments of a long approach.

Acceptable crosswind handling characteristics can be obtained through various design features, practically all of which are incorporated in current production aircraft as listed below. Crosswind takeoffs and landings can be readily performed by the average private pilot if the airplane design has:

1 tricycle landing gear;

2 aileron, rudder, and elevator systems that are responsive at low speed;

3 a combination of wing area and flaps adequate to provide low landing and takeoff speeds (particularly important for short-field and seaplane operation);

4 a responsive engine and throttle system for control in gusty air or for climb-out when a landing must be aborted; and

5 basically stable flight characteristics.

This chapter has covered why these features are important and what design parameters are required to make them effective. To satisfy these criteria, the most popular aircraft look somewhat alike. For the near future, the modern private airplane will have a tapered wing, dorsal fin, effective ailerons, and tricycle landing gear with a relatively free-to-caster nosewheel. Interestingly enough, the design of all these components have been influenced by crosswind operation requirements and related FAA flight demonstration criteria.

Endnote

[1]Fred E. Weick and Richard W. Noyes, "Wind Tunnel Research Comparing Lateral Control Devices, Particularly at High Angles of Attack II—Slotted Ailerons and Frise Ailerons." *NACA Report No. 422*, 1932.

6

Takeoff
and climb

Chapter 5 covered how the primary control surfaces interact to provide acceptable crosswind handling characteristics. In similar, but more complex fashion, the powerplant, wing area, control surfaces, fuselage, landing gear, and design weight must be optimized and combined to provide the best possible performance during all phases of flight. As we get into the fundamentals of airplane design, let us first review the ground handling, takeoff, and climb characteristics necessary for pilot and FAA acceptance of current production aircraft.

Getting underway

While not initially part of the takeoff sequence, ground handling is of prime importance for aircraft acceptance. If an airplane is difficult to taxi away from the ramp or tie-down area, any pilot naturally feels a bit apprehensive and annoyed before the flight begins. This is no way to enjoy flying, or to sell airplanes either, for that matter.

The type of landing gear is most critical, and there is no doubt that a wide tread, tricycle configuration is much easier to handle than the so-called conventional arrangement with main wheels forward plus a tail wheel. The addition of a steerable nosewheel and differential braking complete the features necessary to make tricycle landing gear the easiest to control under all ground handling conditions, including takeoff and landing runs.

In addition to these control benefits, tricycle gear positions the airplane in a fairly level attitude that improves forward visibility, further adding to pilot and passenger safety as well as improved ground handling. It is no wonder then that the most popular late-model aircraft are equipped with tricycle gear, a steerable nosewheel, and independent main wheel brakes.

Taking off

As the takeoff gets underway, there is nothing as satisfying as pushing the throttle home and feeling the seat back give your spine a good shove. This feeling is acceleration, and the amount available can be very critical for operation out of short fields or ponds. Although we seldom experience takeoff acceleration exceeding ⅓ g with any of our small aircraft (for example, 800 lbs of thrust at 2400 lbs takeoff gross weight = 800/2400 = 0.33 g), even this value is difficult to obtain unless the airplane has:

- a constant-speed propeller or a fixed-pitch type propeller designed for maximum climb thrust;
- a weight-to-power ratio of 12 or less (W/hp < 12); and
- relatively low gross weight for the power available.

In analyzing acceleration, the initial takeoff parameter, it becomes clear that propeller design, powerplant size, airframe weight, and useful load are all crucial items. A variation in any one of these affects takeoff distance favorably or unfavorably.

Because thrust is a function of blade design and pitch angle, engine power, ambient temperature, and altitude, our source of accelerating force is obviously sensitive to many things. Reduced thrust delivered by a fixed-pitch propeller in comparison with a constant-speed type means lower acceleration and an increase in time and distance to reach takeoff speed. This occurs because a constant-speed propeller at takeoff has a lower blade angle than a fixed-pitch prop and so delivers greater thrust for increased acceleration. The fixed-pitch propeller blade angle is normally set for cruise flight, which means the blade pitch would be greater than optimum for takeoff. This is necessary because a fixed-pitch propeller designed for maximum thrust at takeoff would overrev at cruising speed due to blade unloading, as explained at greater length in chapter 11. Takeoff time and distance is decreased by replacing a fixed-pitch cruise propeller with a constant-speed one of correct design.

Increased thrust and improved takeoff acceleration comes from added horsepower, provided the propeller is changed or modified to accept increased engine torque. Again we have a trade-off, however, because more powerful engines usually are heavier engines and either the gross weight increases, which tends to lower acceleration, or useful load must be reduced to maintain the original gross weight. In addition, airplane cost and fuel consumption both increase. Conversely, more weight at constant power reduces acceleration, as shown by this simple formula:

$$\text{acceleration} = \frac{\text{thrust}}{\text{gross weight}}$$

This means airframe weight and useful load have a direct bearing on takeoff time and distance, and explains why you must always be aware of takeoff weight vs. airplane design gross weight. Overloading can mean flying into the trees at the end of the runway instead of over them.

Temperature and altitude conditions can also result in a tree-studded take-off. In fact, temperature and altitude combine to form *density altitude*, which has direct effect on engine power, takeoff thrust, and takeoff speed over the ground. An airplane takes off (or stalls) at the same indicated airspeed, regardless of air density, because lift includes the dynamic pressure relationship:

$$q = \rho \ \frac{V^2}{2}$$

where q = dynamic or impact pressure as read by the Pitot tube, in psf
ρ = air density (affected by altitude and temperature), pg. 155
V = airplane velocity in fps

For a given wing area, airfoil section, and angle of attack, the lift developed depends on the mass flow of air passing over the wings. This mass flow must always have the same total value regardless of altitude or temperature when the same load is to be lifted by the same airplane.

It follows from the preceding formula that, as air density decreases, airplane velocity increases to keep q constant. Simply interpreted, as air density decreases, takeoff ground speed and distance increases.

Of course, engine and propeller output also decrease with reduced air density. The additive combination of lower power (which means lower acceleration) and higher ground speed necessary to develop lift can easily exceed the takeoff distance available, again ending the takeoff in the trees. But this time it is due to density altitude rather than excess weight.

For pilot convenience, the effects of temperature and altitude on takeoff and climb have been put into a simple circular calculator known as the Denalt performance computer. It might sound as though this aid was developed by a French pilot or engineer, but the name is actually a combination of "density altitude" contracted into the word Denalt. Separate Denalts are available for fixed-pitch and for variable-pitch propellers, and both can be obtained from the Superintendent of Documents, U.S. Government Printing Office, Washington, DC 20402.

The Denalt performance computer includes takeoff distance variations with density altitude as well as climb rate, and should be part of your flight kit. Because takeoff distance and climb rate are important safety considerations, note that on a 60°F day an airplane equipped with a fixed-pitch propeller requires 1.9 times the takeoff run at 4000-ft pressure altitude than it does at sea level. And if the climb rate was 700 ft/min, at sea level it decreases to 460 ft/min at 4000 ft and 60°F. At 90°F the 4000-ft altitude takeoff becomes 2.1 times the length at sea level, and the climb further reduces to 365 ft/min. So beware of high temperatures and altitudes at takeoff. Not only is the ground run increased, but climb-out over local obstacles is severely decreased as well.

Terrain can affect takeoff as much as density altitude. If we use a smooth asphalt or concrete runway as a standard of comparison, anything else is downhill. For example, a dry, grassy runway increases takeoff and 50-ft obstacle clearance distances by 7 to 10%, while long grass nearly as tall as hay restricts acceleration to such a degree that takeoff is impossible.

Snow and slush are also culprits. In addition to making runways slippery and dangerous, these conditions can freeze wheels and brakes and cause jet flameouts from slush ingested during takeoff. Airline flight standards presently prohibit taking off from runways covered with 2 in. or more of slush. Consider the size of their wheels and thrust-to-weight ratios compared with our small private aircraft. If they have trouble, light aircraft fare even worse.

Blown snowdrifts and hidden ice ridges can be as dangerous to a landplane on takeoff as a submerged sandbar is to a seaplane. There is no design solution for these conditions, nor is there useful advice for that matter, other than recommending the installation of snow skis in place of wheels.

By this time, it must be quite apparent that density altitude and terrain effects can combine into a potentially dangerous set of takeoff conditions. This is particularly true for operation from grassy or snow-covered upland meadows, and partially explains why this type of flying requires experience. The other part of the explanation can be found in local wind conditions due to mountain ridges and slopes, uphill takeoff patterns necessary to clear terrain, and local obstacles that create turbulent flow.

Aircraft with conventional landing gear are best suited for these conditions because the tail wheel can be raised early in the takeoff run and the airplane then held at a trim angle, producing minimum drag during takeoff. The Piper SuperCub is a classic example of such design and short-field capability.

Unfortunately, the relatively short wheelbase of tricycle gear aircraft causes them to hobbyhorse over uneven ground as the nosewheel surveys every contour. This results in a constantly changing wing angle of attack, preventing the buildup of steadily increasing wing lift. Because of the location of the main wheels on tricycle designs, considerable airspeed is required before the horizontal tail can develop enough down load to rotate and lift the nosewheel clear of the ground, as shown in Fig. 6-1a.

This down load is added to the wing load, as was discussed in chapter 2 and Fig. 2-4. Of course, this increases takeoff speed because more wing lift must be developed. The speed required to lift a tail wheel is usually lower than nosewheel lift-off speed because a tail wheel is normally more lightly loaded than a nosewheel. In addition, the distance between the main gear and tail is greater, requiring less tail load to rotate a given gross weight, as indicated by Fig. 6-1b.

Also note that this lift is directed upward, reducing, rather than adding to, wing lift required for takeoff. And this is a primary reason why tail-wheel aircraft are preferred for short-field and bush-country operation; due to their landing gear arrangement, tail-wheel aircraft are able to take off at lower speed and in

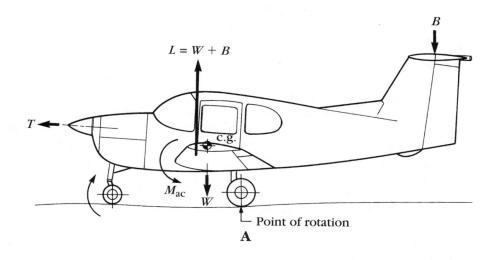

$$L = W + B$$

T

c.g.

M_{ac}

W

Point of rotation

A

$$L = W - B$$

M_{ac}

B

T

c.g.

W

Point of rotation

B

6-1 *Tail loadings required to raise auxiliary wheels clear of terrain. (a) Horizontal tail action required to raise a tricycle gear nosewheel adds to total wing loading. (b) Horizontal tail action required to raise a conventional gear tailwheel reduces total wing loading.*

less distance than tricycle gear aircraft. Once flying speed is reached it is necessary to rotate the airplane about the main wheels to increase angle of attack and so develop the lift required to unstick and climb out of ground effect. This trim change results in a tail down load for tricycle gear aircraft and requires lowering the aft fuselage of tail-wheel types. Both actions tend to increase the wing loading, and to some extent prolong the takeoff run.

In an effort to minimize this effect for tricycle gear aircraft, some designers set the wing at such an angle to the ground line that the airplane can effectively fly off in a three-point attitude. This goal can only be realized on a smooth runway as previously noted. In addition, the three-point attitude generally develops more drag due to increased projected frontal area, so in reality the takeoff run might be increased.

The Cherokee series approaches this three-point arrangement, which has the further objection of causing nosewheel contact damage when the landing speed is too high. The accompanying nose-low flight attitude (frequently referred to as *wheelbarrowing*) can subject the nosewheel to overloading and structural failure.

Of course, tail-wheel aircraft are designed to stall in the three-point attitude and can take off in that fashion, although the added frontal area drag can again be expected to increase the run beyond that of an early "tail up" takeoff technique.

Due to hydrodynamic suction effects, breakout speed is usually well above stall speed for hull seaplanes, and, to a lesser degree, for floatplanes. This explains why a seaplane seems to pop up out of the water once it is ready to take off; the surplus speed is converted into a short burst of excess lift at the takeoff trim attitude.

With landplane, seaplane, tricycle, or conventional (tail-wheel) gear, takeoff speed should be at or near 1.3 to 1.4 times the stalling speed, with the higher value preferred for rough air conditions. As you might know, current FAA design specifications prohibit a gross weight stalling speed in excess of 61 knots (70 mph) for private single-engine aircraft.[1]

Fortunately, most small airplanes stall around 60 mph (52 knots), so our takeoff speed should never have to exceed 85 mph (73 knots), even in turbulent conditions.

As a note of interest, FAR 23 used to require demonstration of elevator power sufficient to lift a tail wheel at or before 0.80 V_{s1} or a nosewheel at or before reaching 0.85 V_{s1} where V_{s1} is the stalling speed in the takeoff configuration. In order to properly handle rough terrain it should be possible to lift either auxiliary wheel at a considerably lower airspeed, and most aircraft can do this quite handily. This former paragraph 23.51 has recently been replaced with a requirement to determine takeoff distance over a 50 ft obstacle for data presentation in the Airplane Flight Manual (AFM).

Improving takeoff

So much for the fundamentals of takeoff, which are basically determined by the simplest laws of physics. Fortunately, we have developed ways to improve takeoff performance beyond these limitations by decreasing the takeoff distance while also increasing the rate of climb. Increased horsepower does this, but at the price of useful load, greater cost, and reduced economy.

A much more effective way to improve takeoff performance is the sound engineering approach of solving a problem at its source. In this case, by lowering the takeoff speed. The lower the end speed necessary to rotate and climb out, the shorter the takeoff run for a given power and weight condition. This is so because if takeoff velocity is reduced but acceleration remains unchanged, the distance required to reach this lower speed is also reduced. In addition, lower takeoff speed means the available acceleration is applied for a shorter time; and so we have a double gain in that both time and distance are reduced proportionately with takeoff speed. And I'm sure you are pleased to know this now completes our takeoff physics lessons.

Of course, lower takeoff speed permits operation from shorter runways and smaller ponds—or safer operation from larger ones, and with a steeper climb angle as well. Reduction of takeoff (and landing) speed is particularly important for seaplane design because hull strength requirements are based on a function of stalling speed squared. As an illustration, reducing the stalling speed 8% from 65 knots to 60 knots lowers the hull bottom loading by 17%.

Three ways to reduce takeoff speed for a given gross weight are:

1 Increase the wing area.

2 Use a more advanced airfoil section having higher lift coefficient values.

3 Use of wing flaps.

None of these provides an ideal solution to the problem in itself. As you review each method, the necessary trade-offs provide an excellent lesson in basic engineering, for practically all phases of aircraft design require compromises favoring operational requirements. In this case, successful compromise can mean the use of all three methods to improve takeoff and landing performance.

The use of increased wing area reduces landing speed in inverse ratio to the square root of the wing area change, as shown by the formula

$$V_{S_2} = V_{S_1} \sqrt{\frac{S_{W_1}}{S_{W_2}}}$$

This means that, for a given gross weight and airfoil section, doubling wing area would decrease an 80-mph stalling speed down to only 56.6 mph—not 40 mph. As might be expected, such wing area increases change more than just the wing weight because more tail area is necessary to satisfy stability requirements; the fuselage weight goes up to carry the greater loads, etc. In fact, the remaining useful load soon drops below an acceptable value.

In practical designs, an increase in wing area can be used to provide some reduction in stall and landing speeds, but if considerable change is required, the designer must seek other solutions.

One solution is an increase in lift coefficient. If the calculated stalling speed is quite high because low lift airfoil was selected, there are many other wing sections available that can deliver increased lift without any adverse side effects. You might tend to then select a very high lift airfoil, but remember that any accompanying high section moment coefficient means an increase in tail loading and structural weight (as shown in chapter 2, Level Flight Loads). This should be avoided.

For current production private aircraft, a basic lift coefficient value approaching 1.3 to 1.4 is not unusual, as shown in Table 6-1. While the newer GA(W)-1 and GA(W)-2 airfoils developed by NASA can exceed these lift values with no drag penalty at cruising speed, their moment coefficients are as much as three times those of the older laminar-flow airfoils and more than 11 times that of the 23015 section.

In final analysis, the airfoil selected is usually a compromise among the structural considerations of maximum thickness and beam location; section drag at cruise; the lift coefficient the designer would like to have; and the tail load the airplane can carry without penalizing either structural weight or performance. For those of you interested in further airfoil study or data, *NACA Technical Report* "Airfoil Section Data Obtained in the NACA Variable-Density Tunnel" is helpful.[2]

Wing flaps are the third way to increase total lift, and flaps are capable of reducing stall and landing speeds beyond the limits possible with wing area increase or through selection of a different airfoil. As shown by Fig. 6-2, flaps come in many shapes and applications, but the type most commonly used on small aircraft is the slotted flap.

Although the Fowler flap provides a greater increase in lift coefficient, as plotted on Fig. 6-3, the rather complex and heavy linkage necessary to extend this type of flap behind the wing trailing edge is too expensive for the added gain. Actually, Fowler flap lift is realized by an increase in both lift coefficient and total wing area resulting from the addition of the extended flap planform to the basic wing area.

Slotted flaps are usually hinged at a fixed point below the flap lower surface, as can be seen on the Cherokee and other aircraft. The Cessna line uses an adaptation or combination of the slotted and Fowler types that is track-mounted but does not go as far aft as a true Fowler flap, and, in fact, behaves like a slotted flap.

The comparative gain in lift coefficient ΔC_L possible with these various flap arrangements is given by Table 6-2, along with the drag coefficient increases and section moment coefficients for flaps lowered 45°. This is about the maximum angle ever needed for small aircraft because flap drag builds up rapidly beyond 40° deflection.

Table 6-1 Airfoil comparison data (see Endnote 2)

Airfoil section	$C_{L_{MAX}}$ @RN = 3 × 10⁶ *	Wing section angle of attack @ $C_{L_{MAX}}$ ∝	$C_{M_{ac}}$ Wing section moment coeff.	$C_{do_{min.}}$ Wing section drag coeff @ std. roughness	Cruise C_L range @ $C_{domin.}$	Percent thickness = $\dfrac{\text{Thickness}}{\text{Chord length}}$ × 100
4415	1.42	12°	−0.092	0.0115	+0.2 to +0.5	15% thick. Gentle Stall. Good lift at altitude.
23015	1.48	15°	−0.008	0.0102	0 to +0.3	15% thick sharp stall
63₂–215	1.41	14°	−0.035	0.010	−0.1 to +0.4	15% thick @ .35 chord laminar flow-fair stall
64₂A215	1.35	15°	−0.035	0.0102	0 to +0.4	15% thick @ .35 chord laminar flow-gentle stall
65₂–215	1.41	16°	−0.035	0.0104	−0.1 to +0.3	15% thick @ .40 chord laminar flow-good stall
65₂–415	1.44	16°	−0.061	0.0102	−0.2 to +0.6	15% thick @ .40 C-laminar Cherokee section-gentle stall
66₂–215	1.41	16°	−0.035	0.010	−0.1 to +0.3	15% thick @ .45 C-laminar good stall
GA(W)-1	1.52	15.5°	−0.09	0.012	0 to +0.8	17% thick @ .40 C-gentle stall-new NASA section
GA(W)-2	1.62	16°	−0.105	0.011	0 to +0.6	13% thick @ .40 C-good stall-latest NASA section

*RN = Reynolds number = 9360 × V_{MPH} × wing chord in feet (an empirical comparison value for flight speed and wing chord).

∝ = Angle of attack =

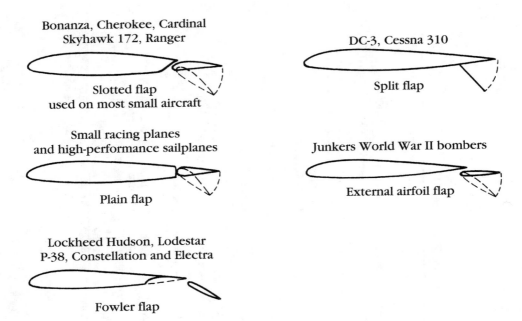

Bonanza, Cherokee, Cardinal
Skyhawk 172, Ranger

Slotted flap
used on most small aircraft

DC-3, Cessna 310

Split flap

Small racing planes
and high-performance sailplanes

Plain flap

Junkers World War II bombers

External airfoil flap

Lockheed Hudson, Lodestar
P-38, Constellation and Electra

Fowler flap

6-2 *NACA 23012 airfoil with different types of flap.*

You might have thought wing design was a simple process; but much compromise is involved in the design, starting with takeoff speed and gross weight requirements. By using a balanced combination of wing area, airfoil lift characteristics, and a trailing edge flap capable of nearly doubling lift over the flapped portion of the wing, the designer can offer a landing speed less than 60 mph with flaps down. This, of course, also means a lower takeoff speed.

As any owner's manual indicates, practically all modern private aircraft equal or better this touchdown speed. Too much wing area means a reduction in cruising speed due to all that surface drag, so of course our modern airplane is equipped with a version of slotted flaps running from the fuselage out to the ailerons. This addition is necessary to improve short-field takeoff and landing control, reduce the stalling speed, and meet FAA maximum permissible stalling speed requirements.

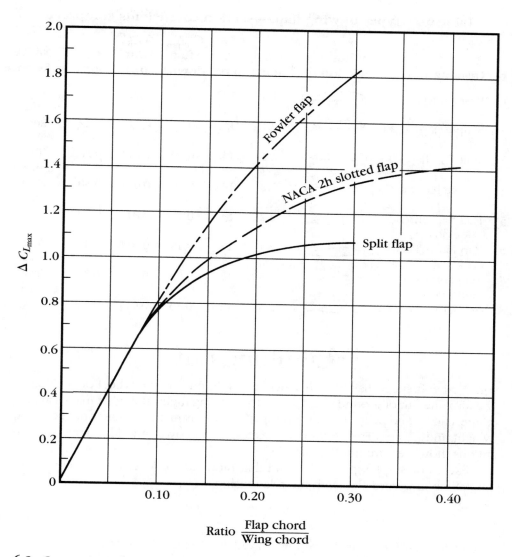

6-3 *Comparison data, NACA 23012 airfoil section with various types of flaps— increase in ΔC_L vs. flap cord.*

Table 6-2 Types of wing flaps—section coefficients compared

No.	Designation	Wing section	C_L max.	$\alpha°$ at C_L max.	C^1_{dmin} (flap down)	C_{dmin} (flap up)	C_M ac	NACA reference
1	Basic airfoil		1.4	17	....	0.010	0.07	Rept. 352
2	Simple flap on M6		2.1	16	0.07	0.011	0.22	Rept. 260
3	Split T.E. flap *CY*		2.1	14	0.19	0.011	0.26	T.N. 422
4	Simple flap on *CY*		2.2	13	0.15	0.011	0.30	Rept. 427
5	Hall flap on *CY*, slot closed		2.2	13	0.17	0.011	0.29	T.N. 417
6	Zap flap *CY*		2.4	14	0.19	0.011	0.40	T.N. 422
7	Symmetrical T.E. airfoil on *CY*		2.5	12	0.06	0.013	0.34	T.N. 524
8	Fowler flap on *CY*		2.8	12	0.15	0.011	0.80	T.N. 419

Climbing out

Once takeoff speed has been reached, a gentle pull on the wheel or stick takes the airplane out of ground effect and into climbing flight. But what makes an airplane climb, and why do some designs climb faster than others of the same weight and power? The answers are excess horsepower and less drag. The solutions, however, are not that simple.

Excess or surplus power beyond that required for level flight is converted into climb in accordance with the relationship

$$\text{Rate of climb} = \frac{\text{thp avail.} \times 33,000}{W} \text{ (in ft/min)}$$

where *thp avail.* = thrust horsepower available (excess horsepower × propeller efficiency)

W = airplane weight in lbs

The problem the designer faces during this phase of flight operation is obtaining maximum thrust horsepower in combination with minimum airplane drag. To see just what is happening, refer to the classic performance diagram of Fig. 6-4, considering only climb-related details at this time.

The upper line is the power-available curve representing the net thrust horsepower vs. the airspeed that is left to propel the airplane after accessory and propeller efficiency power losses are deducted from engine-rated horsepower.

The lower line is the thrust horsepower required by the airplane to maintain level flight at different airspeeds between stall and maximum horizontal flight speed. These curves are usually developed for the gross weight condition at sea level, standard day temperature and pressure.

At some speed point, or possibly for a narrow range of airspeeds, the gap between the power-required and power-available curves will have maximum value; this is the speed for maximum rate of climb. In Fig. 6-4, maximum climb occurs at speed F, as indicated by the maximum gap measured between points G and H, and extends over a narrow airspeed range of similar excess power. For an actual airplane there can be two or three power-required curves for various flap extensions or other configuration changes; but, regardless, at any given air-speed the distance between the two curves is the power available for climb for the particular airplane configuration shown by the power-required curve.

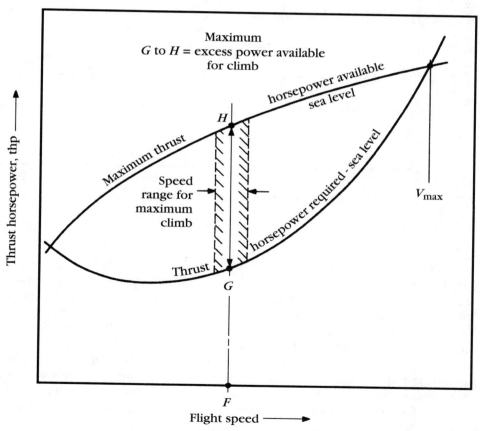

6-4 *Aircraft performance diagram emphasizing climb parameters.*

You can demonstrate this quite easily in your own airplane by following the procedure documented in Fig. 6-5. This is actually a magnified portion of the performance diagram based on flight test data obtained with a 150-hp Cherokee. When this airplane was trimmed at a level flight speed of 80 mph, and using full throttle power with no change in trim setting, the plane climbed at a steady rate of 640 ft/min. Similarly, when the airplane was trimmed at a level flight cruising speed of 120 mph, applying full throttle without touching the trim or elevator controls produced a steady climb rate of about 330 ft/min.

6-5 *Full power climb from cruise trim condition.*

The rate of climb was lower at the higher airspeed because a greater percentage of the total power was being used to obtain a faster cruising speed. As a result, less power was available to provide climb. This can be seen by the decreasing gap between the power-required and power-available curves of Fig. 6-4 as flight speed increases. But regardless, both rate of climb and, therefore, altitude are being controlled by the amount of excess power available.

Obviously, the more we can raise the power-available curve and/or lower the power-required curve, the more thrust horsepower for climb. Neither of these goals is easy to achieve. Just adding horsepower to increase climb might be acceptable for specialized aircraft, but this approach is not ideal for production designs because empty weight, initial cost, and fuel consumption all increase.

Improved propeller efficiency, however, does provide real gain at little or no weight penalty because thrust horsepower is a function of engine power and propeller efficiency. The efficiency of an installed propeller rarely exceeds 75% because of cowling blockage and similar losses, so an efficiency increase of 2.50% represents a power increase of 77.50/75 or a bit greater than 3%. This is a real gain as all the added power can be used to increase climb. Such increases in efficiency can be achieved through improved blade design or refined cowling development.

Just as a constant-speed propeller improves takeoff thrust, it also increases climb rate by providing more efficient blade angle settings at climb speed. A fixed-blade angle propeller set for maximum climb performance unloads and overspeeds in cruising flight, much like the fixed-pitch takeoff prop that was discussed earlier. So a constant-speed propeller is desirable for maximum takeoff and climb performance without any sacrifice of cruising speed. Of course, some penalty is incurred because this type is both heavier and more expensive than wooden or forged aluminum alloy fixed-pitch propellers.

If maximum climb rate, rather than cruising speed, is the design requirement, as would be the case for a towplane used as a sailplane tug, the fixed-pitch propeller is acceptable and preferred; but such applications are in the minority and represent specialized design requirements.

Another, and increasingly frequent, means of obtaining more power is by turbocharging or impeller-blowing the engine intake air. While these forms of supercharging are described in greater detail in chapter 11, the basic principle consists of compressing the induction air sufficiently to develop full rated power from sea level up to some cruising altitude, such as 12,000 to 18,000 ft. Since basic induction system losses regularly steal 5 to 10% of rated power at sea level, supercharging provides added climb power from sea level right on up through altitude cruise. Of course the availability of full power at altitude is a major safety feature during operation from high altitude airports and lakes, to say nothing of clearing the Western Rockies and similar mountain ranges in Canada, South America, and Europe.

Outside of these three ways to increase the available climb power (using a more powerful engine, developing a more efficient propeller, and turbocharging or blowing the induction air), there is little more a designer can do to increase thrust horsepower other than eliminate all accessory drives. This means that for a given powerplant installation, any increase in climb has to be obtained by reducing the amount of thrust horsepower required.

Increasing climb

Because the required-power curve is based primarily on airplane drag and, to a lesser extent, airplane weight, a reduction in either or both of these values reduces the power necessary to maintain level flight, and also increase the rate of climb for a given airspeed.

The climb formula discussed earlier on page 88 indicates that rate of climb (ROC) varies inversely with weight. For example, the ROC increases in direct ratio to weight reduction. This means that cutting the climb weight in half doubles ROC at a given speed and power setting. I think this clearly shows how overloading an airplane can lead to reduced climb, particularly on a hot day—and still worse, if the day is hot and the altitude high. One of the best ways to climb fast is to climb light, a fact we all know but sometimes forget to apply.

Given a fixed gross weight and powerplant installation, the most efficient and permanent method of increasing climb is through drag reduction. In addition to improving flight safety and operating economy, the attack on drag reduction has the capability of being launched from many fronts.

The first, largest, and most effective drag reduction source is that of proper wing design. As wing area is increased at a given flight weight, the lift coefficient required is reduced proportionately in accordance with the basic formula:

$$L = C_L \times \frac{\rho}{2} \times S_W \times (V)^2$$

where L = lift (airplane weight in lbs)
C_L = lift coefficient
ρ = mass density of air = 0.002378 at sea level
S_W = wing area in sq ft (including fuselage area as shown in Fig. 6-6)
V = airplane speed in ft/sec

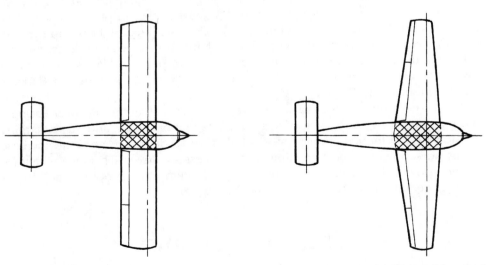

6-6 *Total wing area diagrams. (The shaded areas are considered part of the total wing area.)*

As you can see, for given values of air density ρ, airplane weight L, and airspeed V, C_L reduces in direct ratio to increased S_w. This reduced C_L value, in turn, lowers the wing induced drag C_{D_i}. Reducing induced drag is important because this type of drag is a considerable portion of the total airplane drag at climb speeds, acting in accordance with

$$C_{D_i} = \frac{(C_L)^2}{\pi \times AR}$$

where C_{D_i} = induced drag coefficient

C_L = wing lift coefficient required at the airplane weight and flight speed

π = 3.14

AR = wing aspect ratio = $\dfrac{(\text{span})^2}{\text{area}}$

This formula tells us that cutting the lift coefficient in half reduces the induced drag coefficient by 75%. However, a reduction of this magnitude is not practical because the required wing area increase would be prohibitive, as previously noted. Although some increase in wing area should be used to lower induced drag to an acceptable level, the added wing surface area increases skin friction drag and airframe weight, both of which tend to lower climb rate. So again the designer is faced with a trade-off compromise: this time between desired wing area vs. resulting structural weight and skin friction drag increases.

Referring again to the induced drag formula, notice that increased aspect ratio also acts to lower C_{D_i}, although not to the same degree as reduced C_L. Because increased aspect ratio has a favorable effect on cruising flight drag, it is a desirable design feature, provided any related structural weight increases can be kept under control.

The effect of aspect ratio on induced drag for different lift coefficients is presented in Fig. 6-7, which is a graphic presentation of the induced drag formula. Note that at a cruising C_L of 0.50, the induced drag coefficient is 0.02 at an aspect ratio of 4, but only 0.004 at AR of 20. Five times the aspect ratio results in one-fifth the induced drag. Since an aspect ratio of 20 is too high to be practical for general aviation use from both structural weight and ground handling considerations, some compromise is necessary; possibly finalizing the design at an aspect ratio value of 8 with accompanying induced drag coefficient of 0.01.

It is also apparent that high aspect ratio is not important to aircraft designed primarily for high-speed operation at low values of C_L. At a speed requiring a C_L of only 0.25, induced drag changes vary slightly whether the wing aspect ratio is 4, 8, or 20. However, general aviation aircraft, even those designed for high-speed cruising performance at low C_L values, have to climb to operating altitudes at speeds below cruise. At the comparatively high lift coefficients required during climb, increasing the wing aspect ratio lowers the induced drag, which in turn releases horsepower to increase the climb rate.

6-7 *Induced drag* C_Di *variation with lift* C_L *and aspect ratio AR.*
(AR correction factor = 1.0)

Therefore, for a given flight weight, wing area, powerplant installation, and similar general configuration, it can be stated that the airplane with the greatest aspect ratio has the highest rate of climb. Again some trade-off is necessary because wing structural weight increases with aspect ratio. This happens because the wing root and inboard section bending moments increase as the center of lift moves outboard with increased wingspan. If the wing weight is allowed to get too high, the loss in ROC due to weight increase could exceed any drag reduction realized through increased aspect ratio.

The reduction of interference drag is another important improvement that favorably affects both climb and cruise performance. This type of drag is caused by the breakdown of airflow over external structural and accessory items such as landing gear struts and wheels, skis, wing and tail brace struts, wires, exposed bolt heads and nuts, drain fittings, radio antennas, and similar appendages. In fact, the reduction of interference drag probably represents your maximum opportunity for performance improvement.

In the landing gear area alone, design comparison studies have shown that adding wheel fairings to tricycle gear aircraft improves climb performance midway to that of fully retracted gear. Referring to Fig. 6-8, if a design climbs 750

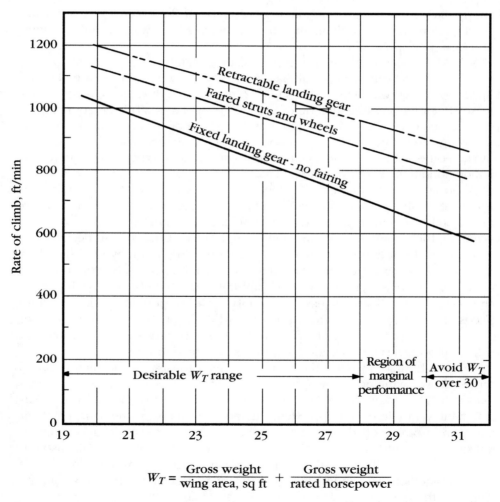

$$W_T = \frac{\text{Gross weight}}{\text{wing area, sq ft}} + \frac{\text{Gross weight}}{\text{rated horsepower}}$$

6-8 *Landplane (tricycle landing gear) climb rate vs. landing gear fairing and* W_T.

ft/min with unfaired tricycle gear lowered, and 1000 ft/min with the gear re-tracted, the fixed-gear climb performance will be increased to about 900 ft/min by adding wheel fairings to all three legs. Faired gear can reduce the perfor-mance attractiveness of retractable landing gear to such a level that the compli-cation hardly seems justified for most general aviation designs, so don't leave your wheel fairings behind, particularly in hot weather when climb becomes in-creasingly critical.

Further drag reduction is possible by sealing cowl and cabin leaks, not leav-ing the cowl flaps fully open on cold days, and keeping the entire airplane pol-ished and free of mud and bugs. You can also streamline or use buried antennas, keep the wing and tail surface leading edges free of dents, and fillet wing and canopy intersections. You can take care of many of these items yourself; each one slightly reduces total drag, and so contributes a bit toward improved climb and cruise performance.

In addition to the effects of weight and drag, climb is also affected by den-sity altitude—that annoying combination of temperature and pressure altitude. Figure 6-9 (based on FAA CAM 8, Fig. 9-2) shows that ROC decreases with in-creases in either pressure altitude or ambient temperature. This deterioration of performance occurs because engine power and propeller thrust decrease with increasing temperature and/or altitude, both of which reduce air density. As with takeoff distance, consult a Denalt computer for ROC when temperature and altitude conditions vary from the familiar pattern.

Although FAA requirements accept climb rates as low as 300 ft/min, de-pending on stalling speed in takeoff configurations, modern aircraft climb in ex-cess of 600 ft/min at gross weight and sea-level standard conditions.[3] For safe operation, this should be the minimum acceptable ROC for aircraft design.

Flaps might or might not increase ROC depending on the type of flap, flap extension, wing area, configuration drag, and climb speed. Certainly the drag producing split flap increases neither climb rate nor angle of climb, while most other types can at least increase the climb angle by permitting slower climb speeds at about the same ROC. In this regard, the slotted flap and versions of the Fowler wing area increasing flap are the best types for short-field climb out over local obstacles and hillsides, as discussed in takeoff design.

While I hesitate to enter the arena, we need to discuss the highly touted "turned down" wing tips at this point. Personally conducted flight test investiga-tion of such tip additions to the Teal amphibian showed slightly improved lat-eral stability in climb and approach, but no measurable increase in climb or reduction in power off descent rate. However, on a high aspect ratio competi-tion sailplane, sculptured wing tips did reduce the stalling speed while also im-proving lateral stability to a slight degree. For production powered aircraft, it is doubtful whether the gain in climb performance exceeds that realized from the wing area these larger tips add to the basic planform.

6-9 *Rate of climb correction for altitude and temperature.*

During the mid 1980s, I conducted some wing tip configuration studies in the Princeton smoke tunnel. The various test shapes included contoured, blown, and induced flow designs, with the most significant being a normal rounded forward tip section that faired into a quarter round upturned trailing edge, as shown in Fig. 6-10. The upturned circular section controls wing tip vortex flow by moving the vortex center outboard of the tip, effectively increasing wing aspect ratio for a given wing span.

This, in turn, reduces induced drag, which, as previously discussed, improves climb and cruise performance. This tip configuration has worked well on

6-10
Upturned circular trailing edge wing tip on the TA16 Seafire Amphibian.

my new SEAFIRE Amphibian and on two other recent designs that have flown successfully. It has also been noted that since the article I wrote entitled *Why Wings Work* appeared in 1986, a number of homebuilt aircraft are flying with this radiused, upturned trailing edge wing tip configuration.[4]

So much for the many factors affecting ROC. Another characteristic that affects pilot acceptance of an airplane during climb is dynamic stability. As I already discussed, static stability can be ideally demonstrated in flight by stick forces that increase linearly as the airplane is displaced from a set of trimmed conditions. This effect is quite similar to a moored boat righting itself from local waves or from someone moving about on the deck. If this same boat when underway recovers equally well from pitch or roll displacement through a series of decreasing oscillations, it is considered to be dynamically stable; but if the boat tends to roll over or pitch under at speed in a seaway instead of recovering to its original upright trim condition, it is dynamically unstable.

Similarly, if an airplane recovers from displacement in pitch, roll, or yaw caused by local turbulence, overcontrol, or normal flight maneuvers, it is considered dynamically stable. If overcontrol or displacement about any of the three axes results in a continued or increasingly divergent oscillation, the airplane is dynamically unstable. This condition is very undesirable for any private airplane because the pilot must pay constant attention to maintain a heading, desired altitude, or trim condition.

Nothing can be more annoying than an airplane that constantly oscillates about one or more axes during climb, particularly in turbulent air conditions. For this reason current FAA design requirements place heavy emphasis on stability about all three axes (pitch, roll, and yaw) and throughout all phases of flight. This is why your modern production FAA certified airplane does not exhibit unstable tendencies in climb so common to early aircraft of all types.

At some altitude, all thrust horsepower available will be required to keep the airplane in level flight at a speed just above indicated stalling speed. The altitude at which this occurs is known as the absolute ceiling; the airplane cannot climb higher for a given takeoff weight. As another standard of performance, the altitude at which ROC has decreased to 100 ft/min is known as the service ceiling; this is lower than the absolute ceiling and is an altitude at which some maneuvering is possible, provided all controls are well coordinated.

As a final note, sea-level climb rate and service ceiling are important design points for single-engine aircraft, while the single-engine ceiling is a major design consideration for twin-engine aircraft. For all normal aircraft, sea-level climb, service ceiling, and single-engine ceiling are improved by installing either constant-speed propellers or turbocharged engines, or both.

In summary, because maximum ROC occurs at the speed for airplane maximum lift to drag ratio L/D, the lower the total drag the greater the ROC. In addition to this basic drag consideration, ROC is also a function of available power, airplane weight, and density altitude. As you have seen, satisfying any one of these factors affects all the others, requiring considerable design trade-off study to obtain acceptable climb performance within practical horsepower, wing area, useful load, and operating altitude limitations.

Endnotes

[1]FAR 23.49(b).

[2](a) Jacobs and Abbott, "Airfoil Section Data Obtained in the NACA Variable-Density Tunnel." *NACA Technical Report No. 669*, 1939.

(b) Abbott and von Doenhoff, *Theory of Wing Sections—Including a Summary of Airfoil Data.* Dover Publications, Inc., New York, 1959.

[3]FAR 23.65(a).

[4]D.B. Thurston, "Why Wings Work," Part 3. *Kitplanes Magazine*, October 1986, p. 73.

7
Cruising with stability

After we climb to the desired altitude, we begin cruising to our destination, and this is the phase of flight for which most private aircraft are purchased. Whether used for business or personal convenience, modern aircraft offer a great time-saving advantage over the automobile because they can fly between two points in an essentially straight line. The faster the speed along this line for a given powerplant size, the more economical the airplane. These are obvious facts, but the ways to attain them are not so obvious.

What price cruising speed?

Increasing the size of the engine raises the cruising speed, but, as with takeoff and climb, at the sacrifice of useful load and fuel economy, as well as increased initial cost. And the speed increase with increased power is not all that great because, for the same altitude, speed varies in direct ratio to the cube root of the power change, thus

$$V_2 = V_1 \sqrt[3]{\frac{(hp)_2}{(hp)_1}}$$

where V_1 and $(hp)_1$ = the original speed and horsepower
V_2 and $(hp)_2$ = the new speed and increased horsepower

If we cruise at 130 mph with 150 hp, this formula tells us the cruising speed will increase to 143 mph when a 200-hp engine is installed, offering a gain of 10% in cruising speed for an increase of 33% in power and fuel consumption. This is not a particularly exciting trade-off, but it is typical.

For example, the 180-hp Beechcraft Sundowner cruises at 136 mph with fixed tricycle gear, while the 200-hp Beechcraft Sierra of identical wing area cruises at 151 mph with retractable gear and a slightly higher gross weight. These differences are important design considerations, however, because in-

creasing airplane weight with the same wing area means the lift coefficient and
wing drag also increases slightly. Since our formula tells us the 200-hp Sierra
should cruise at 141 mph with fixed gear vs. the published cruise of 151 mph
with retractable gear, we see that by retracting the Sierra landing gear Beechcraft
engineers were able to reduce cruising drag sufficiently to carry the increased
gross weight with some gain in speed; but at the price of increased empty
weight, initial cost, fuel consumption, and maintenance. In view of this, what
practical alternatives do we have available to increase cruising speed and fuel
economy?

To examine this speed problem, let us again refer to the aircraft performance
diagram, now adding the cruising power-available curves to obtain Fig. 7-1. This
figure graphically shows what determines (or limits) flight speed, while also sug-
gesting ways that speed can be increased for more economical operation.

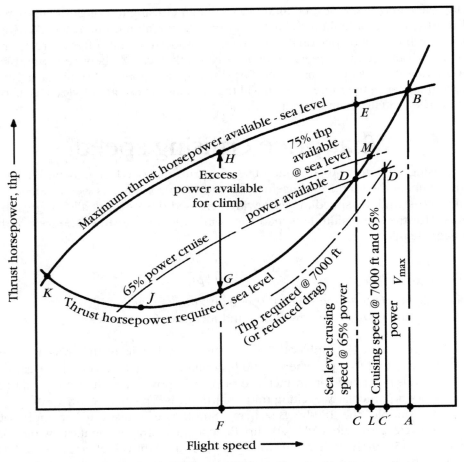

7-1 *Aircraft performance diagram showing speed limits.*

The most effective approach to increase efficiency is through reducing airplane drag, which in turn will reduce the thrust horsepower required at all flight speeds. As you can see, at 65% power our airplane is limited to cruising speed *C* reached at point *D*, which is formed by the intersection of the power-available and the power-required curves. This intersection determines level flight speed for a given altitude and power selection; if there is no more power available at a given throttle setting, the airplane just cannot fly any faster at that altitude.

Assuming these curves are given for sea-level operation, which is the usual procedure, it is possible to decrease cruising drag somewhat by climbing above sea level to regions of reduced air density. Up to the altitude requiring full throttle operation to obtain the power desired, engine cruise power can be maintained at a constant percentage of sea-level power by steadily advancing the throttle. Since airplane drag decreases with air density, the thrust horsepower required to maintain level flight decreases with altitude. This means that as long as the desired percentage of sea-level cruising power can be maintained, cruising speed at that power will increase with altitude up to the flight level requiring full throttle operation.

This explains why an unsupercharged airplane can cruise at increasing airspeed with increasing altitude up to some power-limited speed such as *C'*, determined by point *D'* intersection of Fig. 7-1. For modern private aircraft this point is reached between 6500 to 8000 ft, depending on the powerplant induction system. So if you have a long flight ahead, this is why you should cruise at altitude, weather permitting. For some propeller-driven aircraft, full throttle cruising speed at altitude exceeds full throttle V_{max} at sea level.

All of this shows the obvious need to reduce drag in order to increase cruising speed and airplane efficiency. As one approach to this goal, the designer can reduce frontal area, which usually means a loss of elbow and head room, or increased cabin intimacy à la early Mooney aircraft (which are now improved inside with the same width outside). Sailplanes carry this concept to the *n*th degree by reclining the pilot to such an extent it is a wonder we don't have more accidents caused by dozing pilots. But because the sailplane power source is gravity, the less drag, the less descent required; and by reducing sailplane frontal area and interference drag, the design glide ratio steadily advances toward the magic value of 40 to 1.

As another approach, airplane cruising drag can be reduced by decreasing wing area, but overdoing this reduces climb rate and increases takeoff and landing speed, as discussed in chapter 6. The necessary design trade-off usually requires the use of flaps to obtain landing speeds acceptable to the FAA, plus more wing area than desirable for maximum cruise in order to provide minimum acceptable climb rates.

Although a specialized design type, the early Cleveland Air Racers illustrated this point very well. Their wing area was reduced to such a degree that extremely long takeoff runs and very shallow climb angles were accepted in order to obtain the speeds required from the powerplants available. In fact some of these designs were barely able to get airborne with their design wing area/powerplant combinations.

Thinner wing sections, increased aspect ratio, improved engine-cooling airflow, faired canopy and cowling assemblies, wheel fairings or retractable landing gear, buried antennas, sealed control surfaces, and many similar items all contribute to drag reduction. These possibilities are explained at greater length in chapter 10. At this time, it is sufficient to say that reducing airplane drag lowers the required thrust horsepower. As a comparison, let us consider a reduced-drag power-required curve lying below the sea-level curve values shown on Fig. 7-1. This means that through reduced airplane drag an intersection similar to D' could be realized at sea-level speed C', with higher cruising speeds available at somewhat greater altitudes. Such an improvement provides a decided gain in aircraft performance and efficiency.

To expand this cruising speed discussion a bit further, if the throttle is advanced to 75% power at sea level, we obtain a higher cruising speed L resulting from the intersection at point M of the 75% thrust horsepower available and the basic sea-level power-required curves. In a similar manner, power can be increased right up to point B, resulting in sea-level maximum speed A. Again, this speed would increase with altitude up to some relatively low flight level, depending on the powerplant design and airplane configuration.

Since drag reduction offers improved operating efficiency throughout the entire flight from takeoff to landing, it is important that all airplane surfaces be kept clean and polished and all air leaks sealed to the extent possible. Drag reduction work is time well spent to obtain a safer and more efficient airplane.

Flight stability

Aircraft certified within the past 30 years must satisfy very demanding stability requirements. For example, modern aircraft might not meet certain directional stability criteria by using the rudder; the airplane itself must meet specified directional stability limits without pilot assistance. In addition, if a rudder trim system is used, it must be able to fly the airplane if the primary rudder controls fail, which is a very remote possibility and really a dubious requirement for small aircraft.

Longitudinal stability (about the pitch axis) must be so positive that little trim change or stick force increase is required when executing a missed approach or balked landing with the airplane fully trimmed in the landing configuration. Minimum trim change when applying flaps or power, plus the requirement that an airplane can be flown using the longitudinal trim system (elevator tab or adjustable stabilizer) are other major longitudinal stability criteria; but there are many more.

To complete the stability picture, amendment 14 to FAR 23 added a requirement for minimum acceptable rates of roll during takeoff, as well as in the landing configuration; so now all new private aircraft must meet rigorous minimum stability requirements about all three principal flight axes. In fact, once trimmed, an airplane must be able to fly itself in smooth air. In the remainder of this chapter, we will consider how this level of flight stability can be realized, as well as why it is desirable and important. Once again, the designer is faced with a very interrelated set of configuration problems.

Longitudinal stability and tail power

Lateral and directional stability effects tend to act together. For example, a rather rapid application of rudder raises the outside wing in a turn because of increased lift on the outside panel. In this case, the rudder and fin produced turning lift (directional control and stability), which rotated the airplane about a vertical axis, and so accelerated airflow over the wing panel on the outside of the turn. This accelerated flow produced more lift on the outside panel than on the inner one and this made the airplane bank into the turn (lateral control and stability) by rotating about the fore and aft axis. This is typical of the lateral directional stability relationship. However, longitudinal, or pitch stability, acts quite independently of either of these axes and is the most important of the three.

The primary requirement for longitudinal stability is adequate horizontal tail power. Horizontal tail power is a combination of horizontal tail area and the tail arm: the distance from the airplane center of gravity to the horizontal tail center of lift. The greater the tail arm, the smaller the horizontal tail can be, explaining why sailplanes have long tail arms and small tail surfaces to reduce both tail surface drag and control surface displacement drag.

These same options are available to the powered-aircraft designer, but because powered-aircraft horizontal tail loads are greater than those for sailplanes, some structural weight penalty is incurred when an aft fuselage becomes overly long. On the other hand, when the tail arm is too short, any airplane will tend to become marginally stable longitudinally, and finally unstable in pitch as the center of gravity moves aft.

FAA flight test requirements include a demonstration of acceptable longitudinal stability at a gross weight center of gravity located further aft than the most rearward center-of-gravity limit to be finally approved for certification. Because this aft point is often determined by pilot opinion, unless some dangerous spinning or recovery tendency has been exhibited during flight test, a lively discussion frequently develops between the FAA and the manufacturer's personnel at this point in the flight demonstration program. However, rest assured that if you fly within the aft center-of-gravity limits for your airplane, you will always have positive stability and should never experience a reversal of stick force under any flight condition. But the degree or level of positive longitudinal stability steadily decreases as the center of gravity moves aft because the tail arm is decreasing.

For the benefit of those interested in aircraft design and as a good first approximation, the one-third chord point of the horizontal tail should be located 2.75 to 3 times the wing mean aerodynamic chord (MAC) distance from the wing quarter chord point (or airplane center of gravity).

As we previously discussed in chapter 1, the longer the tail arm the more stable an airplane feels and actually is in flight. The simple reason for this is the longer the tail arm, the more damping action provided by the horizontal tail. To meet FAA design and flight test longitudinal stability requirements, while developing an airplane with acceptably low elevator control forces under all flight conditions, the designer has to consider and optimize the tail arm, the horizontal tail area and lift coefficient, surface drag, structural weight, and the type of horizontal tail design to be used.

Tail selection

We need to discuss tail surface design at some point, and this stability discussion seems a logical spot because the horizontal tail is the principal determinant of longitudinal stability.

An airfoil suitable for tail design will usually be of symmetrical section, and preferably no thicker than 12% of the chord length. Although the tendency is naturally toward selection of a thicker section to reduce bending stresses, the airfoil section lift slope falls off rapidly above 12% thickness while, of course, the section drag increases. The lift slope of an airfoil is based on the straight portion of the lift curve and is the ratio of lift coefficient increase per degree change in angle of attack. The higher the lift slope value, the greater the tail lift and damping action, which in turn means increased longitudinal stability and control response.

For suitable tail surface airfoils, the following shows decreases in lift slope with increased section thickness:[1]

NACA section	Lift slope value
0006	0.1125
0009	0.1094
0012	0.1094
0015	0.0750

If you are interested in the design of horizontal tails providing an acceptable level of longitudinal stability, the following formula permits simple calculation of the minimum surface required in relation to wing area and tail arm:

$$P_{HT} = \frac{D_H}{MAC} \times \frac{S_{HT}}{S_W} \geq 0.55$$

where P_{HT} = horizontal tail power, an empirical value
D_H = distance from center of gravity to 0.30 chord horizontal tail, ft (horizontal tail arm)
MAC = wing mean aerodynamic chord (or average wing chord), ft
S_{HT} = total horizontal tail area, sq ft
S_W = total wing area, sq ft

According to NACA test results, the all-movable and trimming stabilizer configurations require less area than a fixed stabilizer, so the above PHT value could be reduced about 10% for such designs.[2]

Similar to determining an adequate horizontal tail, vertical surface area should equal or exceed the value

$$P_{VT} = \frac{D_V}{\text{MAC}} \times \frac{S_{VT}}{S_W} \geq 0.30$$

where P_{VT} = vertical tail power, an empirical value
D_V = distance from center of gravity to 0.30 chord vertical tail, ft (vertical tail arm)
MAC = wing mean aerodynamic chord (or average wing chord), ft
S_{VT} = total vertical tail area, sq ft
S_W = total wing area, sq ft

NOTE: Flight tests have shown that the value of P_{VT} must be larger for hull seaplanes having considerable side area ahead of the wing leading edge.

To satisfy Colonial Aircraft Skimmer (now Lake) and Thurston Model TA16 Seafire Amphibian directional stability requirements, it was necessary to increase vertical tail area sufficiently to provide a P_{VT} value of 0.42. Also, the addition of floats to a landplane can require increased vertical tail area, either by adding a ventral fin on the aft fuselage, or end plates on the horizontal tail surface.

While the planform design of the horizontal or vertical tail is frequently as much a matter of artistic effort or structural simplicity as aerodynamic consideration, tail aspect ratio should be as high as weight allowances permit, which really means the aspect ratio should be as high as structural loading permits. This is generally around 3 to 4 for horizontal tails and 2 to 3 for vertical surfaces.

To summarize, to be effective aerodynamically and still remain within structural weight limitations, all modern tail surfaces are of fairly thin section and relatively low aspect ratio.

Horizontal tail position on the fuselage or fin is not so typically standardized or well defined. We have low tails, mid tails, T tails, upright V tails, and inverted V tails, all with different effects on damping, control response, control system design, structural weight, and, possibly most important of all, spin recovery effectiveness.

To go back a few generations in design, the designers of the first airplanes not only followed a random approach for vertical location and planform of the horizontal tail, but also frequently placed the tail forward, or aft, or at both locations. Although they soon became leaders in elevator surface and control design, I'm sure the Wright brothers' long periods of tethered glider flight were primarily elevator system design, development, and training flights. They were learning the response of the airplane to elevator trim changes. No one had ever previously documented this phenomenon, nor was there anyone around to teach the required technique. So the Wrights located the elevator ahead of the airplane where it could be closely observed in operation, and learned how to fly. But many builders immediately adopted this tail location because the Wrights were so extremely successful for their time.

Fortunately, as flyable aircraft designs increased in number, Bleriot, Antoinette, REP, and many others realized why an arrow has feathers at the aft end. By 1911 to 1912 most designers had located the horizontal tail behind the wings where it would be most effective at low airspeeds. From the structural design standpoint, the aft and low horizontal tail position was and still is the lightest arrangement because it is the easiest to attach to the fuselage. The earliest aircraft simply tied the fixed portion of the horizontal tail or tails to the fuselage booms, which provided a convenient location for a light surface. Figure 7-2 shows this type of mounting on a 1911 Voisin, apparently designed with one of the first T tails. Early Curtiss aircraft had similar, rather fragile, tail mountings as shown in Fig. 1-2.

7-2 *1911 Voisin tail assembly.*

When higher speeds were realized during World War I, the need to reduce turbulence generated by exposed struts and flying wires, as well as from slip-stream passing over unfaired tail booms, became quite obvious. As a result, fuselages were enclosed around the tail surfaces, and so the low- to mid-position of the horizontal tail developed as the standard location for many years.

When the rudder was not carried down below the horizontal tail, as was frequently the case with early designs, some aircraft experienced extreme difficulty in recovering from either intentional or inadvertent spins; a condition that was caused by horizontal tail blanketing of the rudder, as shown in Fig. 7-3a. This type of blanketing not only prevented rudder reversal from stopping spin rotation, but also frequently contributed to rudder lock in the direction of spin rotation—with fatal results. Subsequent spin studies produced the recommended horizontal tail locations of Figs. 7-3b and 7-3c, as well as the extended rudder shown in Fig. 7-3c.

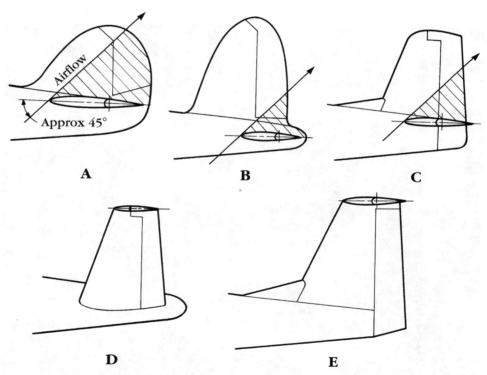

7-3 *The effect of horizontal location on rudder response during spin recovery. (a) An early type of rudder design subject to blanketing by the horizontal tail. (b) Horizontal tail location for improved rudder effectiveness during spin recovery. (c) Rudder extended below the horizontal tail for increased effectiveness during spin recovery. (d) Bonanza type V-tail design provides fully effective tail surfaces during spin recovery (inverted V tail is similar). (e) The T-tail configuration has the fin and rudder completely effective with no blanketing.*

As a further refinement, right after World War II Beechcraft developed the distinctive V tail of the Bonanza series to provide increased horizontal and vertical tail effectiveness throughout the entire flight spectrum. Some drag and cost reduction were also anticipated due to having only two surface assemblies instead of the customary three. However, like some modern medicines, the V-tail configuration has some side effects that are not wholly desirable, as we will discuss later on.

At present, all these horizontal tail locations are gradually being displaced by the T-tail arrangement of Figs. 7-3e and 7-4. While some of the earliest aircraft might have carried a T tail more as a matter of convenience than aerodynamic refinement, Focke-Wulf of Germany developed the first really engineered T-tail design for their late World War II model Ta-183 jet fighter. In fact their initial design even had a swept horizontal tail mounted on top of a swept fin, and with horizontal tail dihedral, as well, to provide the ultimate in T-tail design.

Beech Aircraft Corp.

7-4 *T-tails arrive at Beechcraft.*

The T tail has been slowly accepted for private aircraft because it is generally heavier than a conventional tail assembly due to combined horizontal tail and fin bending loads that must be carried by the fin and fuselage structures. The control system is also more complex and somewhat heavier than for a lower horizontal tail location. There is some gain in directional stability, however, because the end plate effect of the horizontal tail increases fin and rudder effectiveness, while the entire vertical tail acts to assist in spin recovery. With careful design, the increased vertical tail effectiveness can reduce vertical tail area and weight, so some reduction in penalty is possible. If you are interested in T-tail design procedures, a few useful reports are listed in the endnotes.

The action of the various tail arrangements in maneuvering flight and gust conditions seems to be an item of interest to everyone. Briefly, the conventional low horizontal tail is affected by vertical gust conditions, while only the vertical tail responds to horizontal gusts, as shown in Fig. 7-5a. The elevator and rudder act quite independently to provide pitch and yaw control.

There is some increase in surface pressure acting over the horizontal tail when the vertical tail is loaded by rudder displacement or horizontal gust action, as shown by the load arrows of Fig. 7-5a. In addition, downward gusts increase the pressure field acting on both the horizontal and the vertical tails as indicated. Note that such loadings generally act symmetrically on each side of the vertical and horizontal surfaces.

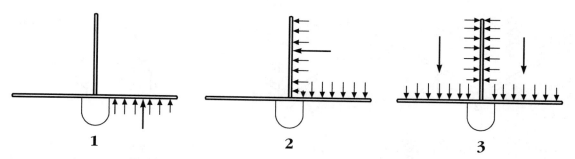

1   2   3

7-5a *Loading condition for the conventional tail surface configuration. Note that airflow can act on either side of the surface and that symmetrical loading of the horizontal tail can occur in an up or down loading direction. (1) Vertical gust or maneuver airflow acting on one side produces unsymmetrical loading on horizontal tail. (2) Horizontal gust or maneuver airflow loads vertical tail and increases pressure over one side of horizontal tail as well, producing unsymmetrical loading of horizontal tail. (3) Vertical down gust or maneuvering airflow acting on both sides increases the horizontal tail load. Balanced loads on the vertical tail result in symmetrical loading of horizontal and vertical tail surfaces.*

It is the horizontal gusts or maneuver-induced airflows which normally develop tail surface unsymmetrical loadings, but these values are usually quite small for the conventional horizontal tail position. For private aircraft, FAA design specifications only require a maximum differential of about 30% in loading between opposite sides of the horizontal tail, including gust and side-slip conditions.[3] In this design condition, one side of the horizontal tail is loaded to 70% maximum design load, while the opposite side is loaded at 100%.

Of course, unsymmetrical tail loading also places a twist, called torsion, on the fuselage aft section. Due to maneuvering or gust action, fuselage torsional loading is frequently further increased by the addition of vertical surface loads; and all these loadings acting together might require a very rugged aft fuselage structure, as we will discuss in chapter 12, Airframe Design.

The action of the V tails is considerably more complex than that of conventional surfaces. As you can see from Fig. 7-5b, the well-known Bonanza upright V tail (Fig. 1-5) acts to simultaneously yaw and pitch the airplane when encountering either vertical or horizontal gust action. This is why the Bonanza seems to yaw with pitch trim changes in turbulent air and vice versa, a condition similar to *hunting*, which is a term applied when an airplane oscillates in flight between two trim points. This characteristic can be nausea producing over prolonged periods in rough air, particularly for anyone who flies infrequently. Could this be the reason more conventional tail surfaces are used on the later Debonair and subsequent Beechcraft designs?

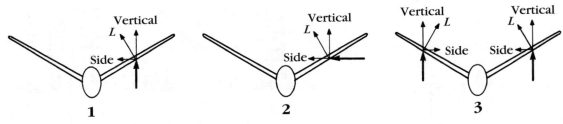

7-5b *Gust and maneuver loading conditions for the upright V-tail configuration. Airflow may act on either side and in an up or down direction. (1) Lift component of vertical airflow produces both vertical and side loadings, moving airplane in pitch and yaw. (2) Lift component of horizontal airflow produces both vertical and side loadings, moving airplane in pitch and yaw. (3) Lift components of symmetrical vertical loading (up or down) act to move the airplane in pitch, but the side components cancel one another as well as any tendency to yaw.*

The inverted V tail of Figs. 7-5c and 7-6 has the same interaction tendencies, but it has a slight design advantage over the upright V tail in that vertical tail area located below the center of side area (approximately at the center of gravity) tends to roll an inverted V-tail airplane into a banked turn. This means any tendency to be lax on rudder action is somewhat self-correcting on aircraft having inverted vertical tails. Because nothing is free in the world of aircraft design, the

7-5c *Inverted V-tail configuration. Lift component of vertical airflow produces both vertical and side loadings, moving the airplane in both pitch and yaw. The side force, however, acts in a direction opposite that of the upright V tail. Loadings are otherwise similar to those of Figure 7-5b.*

7-5d *T-tail configuration under vertical airflow loading. Note pressure increase on vertical tail as well. Control surface deflections also increase loadings similar to gust action; otherwise, load conditions are similar to those for the conventional tail of Figure 7-5a.*

7-6 *Mini Imp homebuilt with inverted V tail.*

inverted tail presents a basic configuration design problem: providing adequate tail clearance for full stall landings. An overly long and heavy landing gear is the usual solution to this requirement, as well as being a reason the inverted V-tail arrangement is not popular.

Other general objections to V tails are that they require more complex and somewhat heavier control systems, plus the heavier tail structure necessary to support combined horizontal and vertical surface loadings. In summary, the V tail is an unconventional arrangement which is generally as heavy as the conventional design it would replace, and the stability characteristics are somewhat less desirable. This is particularly so in rough air, just when a high level of stability becomes increasingly important.

Our final configuration analysis covers the T tail, which is also subject to load interaction to some degree. Figure 7-5d shows how vertical or horizontal gusts and control-surface deflections tend to increase the air pressure acting over T-tail surfaces. Fortunately, these pressure changes are noted more for increasing structural loading rather than producing any adverse stability effects, since a T tail behaves like conventional surfaces.

The interaction of airloads with the T-tail configuration was noted in a recent amendment to FAR 23 requiring combined application of vertical and horizontal tail surface loadings.[4] So although the T tail offers stability advantages without any penalty in landing gear design, a penalty is incurred through increased weight of the fin and aft fuselage structures. This penalty can be somewhat reduced by the increased fin area effectiveness previously noted, but the T-tail structure is still somewhat heavier than a conventional tail arrangement. Also, as was true for the V-tail design, both the elevator control and longitudinal trim systems tend to become somewhat more complex and a bit heavier.

It is also true that unless the horizontal surface of a T tail is located in the slipstream, elevator power will be noticeably absent at low airspeeds, such as the early part of a takeoff run. As a result, a pilot not familiar with the airplane tends to apply excessive up elevator at low speed, followed by sudden pitch-up when aerodynamic speed and control take over.

Such was the case with T-tail Piper Tomahawk, Arrow IV, and Lance II designs, as well as the Beech Model 77 Skipper trainer.

Regardless of these negative trade-offs, the T tail is gaining in popularity because of the improved stability offered for special design applications, such as the minimum trim changes required with increased power for the Teal amphibian (Fig. 4-13). More importantly, it is gaining popularity because of the excellent spin recovery characteristics offered by a completely unblanketed vertical tail.

Since rapid spin recovery is essential to reduce low-altitude fatalities, designers are paying more attention to this most crucial flight condition. Related test results and criteria for desirable spin recovery characteristics are given in articles listed in the endnotes.[5] Although there is little we can do to correct the stall/spin characteristics of existing aircraft, I'm sure that unfavorable accident statistics would continually decrease if everyone spent some time obtaining qualified flight instruction in spin recovery procedures.

Other stability factors

Although the entire subject of flight stability has filled many books, the primary requirements of center-of-gravity location, tail arm, and adequate tail surface design have just been covered in some depth. In addition, some other factors affecting cruising flight stability and handling characteristics should be briefly reviewed for background information. These are dihedral, adverse yaw, thrust line location, and the longitudinal trim system.

Although dihedral and adverse yaw have been discussed in chapter 5, it is important to note that in turbulent air too much or too little dihedral will require constant aileron attention to correct lateral trim. If the aileron design develops any amount of adverse yaw, constant rudder correction will also be necessary in addition to all the other chores facing a pilot in rough air. This is a most annoying situation and an increasingly dangerous one when nearing minimum descent altitude levels during IFR operation. Dihedral is also tied to vertical tail area in developing adequate lateral stability. If the combination is not within certain limits, an airplane can develop a steady rolling and yawing condition referred to as Dutch roll. While there are many explanations for this undesirable flight pattern, the airplane actually goes through a slow oscillation of a 5 to 15° roll on each side of the flight axis, combined at the same time with about a 5 to 10° yaw on each side of the flight path. This behavior is similar to a runaway two-axis autopilot and, as can be imagined, is most disturbing during flight, to say nothing of really blind IFR conditions.

FAA flight test procedures ensure that Dutch roll is not present in recently certified aircraft, but it can exist in some pre-World War II designs when flying in turbulence. This stability problem is usually corrected by reducing the dihedral angle and/or increasing the vertical fin area. Military aircraft frequently resolve this condition by complex autopilot and feedback control system design. Fortunately that expensive approach is not necessary at private aircraft speeds. Instead, our airplanes should have good ailerons, causing minimum adverse yaw, adequate vertical fin area, and not too much dihedral.

Minimum trim change with power is important during cruise because less pilot attention is required to hold altitude or a desired heading. In view of this, the location of the thrust line relative to the center of gravity becomes another important design consideration. When the thrust axis passes far from the center of gravity, trim changes might accompany every change in power setting. Only variations in pitch attitude are normally developed with power changes in single-engine aircraft.

But in twin-engine aircraft, both pitch and yaw trim variations should be expected from power changes where the thrust lines might be outboard from and above or below the center of gravity. Since we discussed basics of this subject in chapter 2 and Figs. 2-4 and 3-4, only a special stability case will be mentioned at this point.

If any design ever violated the principle of having the thrust line pass near the center of gravity, the Teal amphibian of Fig. 4-13 really does. Yet because of the T-tail design and horizontal tail location, changes in slipstream velocity with power result in tail loadings almost exactly offsetting thrust moments about the center of gravity. So the T tail is a definite asset to an aircraft configuration of this type.

Having an airplane slowly hunt or change altitude during cruise is extremely annoying, and generally becomes worse in rough air. Such changes can be partially the result of a poor longitudinal trim system. Whether normally located on the fuselage or in the T-tail position atop the fin, I prefer a fixed or trim adjustable stabilizer to either the all-movable or V-tail designs. My experience has shown that the all-movable horizontal tail is subject to annoying trim changes with slight variations in airflow, and also with slight wear and play in the control and trim system linkages. The V tail causes both vertical and lateral displacement with each gust, as I previously noted. Quite likely the fixed stabilizer is a major reason for the solid feel and resulting popularity of the Cessna line of single-engine 172 Skyhawk aircraft.

The longitudinal trim system connecting the trim wheel with the elevator tab needs to have virtually no cable stretch or linkage play if you are to have excellent positive trim response. We all appreciate a stable airplane that willingly trims to a 20 ft/min adjustment, and it is also a much easier and safer airplane to fly when the going gets a bit tight.

Putting it all together

Stepping back a bit and putting together all these important parameters for cruising flight, we find that besides meeting FAA design and flight demonstration requirements, the design group also had to put together an airplane acceptable to the sales department and owners. To accomplish this for cruising flight requires a design that:

- Has comparatively low drag to provide timesaving, efficient transportation for the powerplant used and the production cost.
- Has ample and positive stability about all three principal flight axes, including pitch, roll, and yaw. This requires the proper relationship and design compromise with the center of gravity, tail arm, horizontal and vertical tail areas, and the dihedral angle. This is demonstrated in flight by return to trim attitude when displaced, positive control forces which steadily increase with control displacement from the trimmed position, and maintaining an assigned altitude and heading in smooth air.
- Has small trim displacement and elevator load changes with power.
- Has an effective, responsive, and positive longitudinal trim control system.
- Probably most important, has a powerplant that keeps operating (as we will cover at length in chapter 11).

Endnotes

[1](a) David B. Thurston, "T Tails—Past, Present, and Future," *Aero Magazine*, May–June 1973

(b) Donald R. Riley, "Effect of Horizontal Tail Span and Vertical Location on the Aerodynamic Characteristics of an Unswept Tail Assembly in Slideslip." NACA TN 2907 (later issued as TR 1171), February 1953.

(c) R. R. Tumlinson and H. L. Walter, "T-Tail Aerodynamics of the Super King Air." *SAE Report 740370*, April 1974.

[2]Sidney M. Harmon, Comparison of Fixed Stabilizer, Adjustable Stabilizer, and All-Movable Horizontal Tails. NACA Report L-195, October 1946.

[3]FAR 23.427 (a) and (b).

[4]FAR 23.427 (c).

[5](a) James S. Bowman, Jr., and Sanger M. Burk, Jr., "Stall/Spin Research Status Report." NASA research contained in *SAE Report 740354*.

(b) James S. Bowman, Jr., "Summary of Spin Technology as related to Light General Aviation Airplanes." NASA TN D-6575, December 1971.

(c) Seth B. Anderson, "Overview of Stall/Spin Characteristics of General Aviation Aircraft." *Sport Aviation*, May 1989, p.19.

8

Approach conditions

Although I didn't list it with primary operational criteria discussed earlier, the approach phase of flight presents some interesting design problems that are worth exploring.

Stability and controllability are all important during approach, and should remain fairly constant over a broad range of relatively low airspeeds (assisted by using flaps, if necessary). As a general rule, the final leg approach speed should be about 1.3 times configuration stalling speed in smooth air, and increased to 1.4 times stalling speed for rough air. But within these limits, the aircraft designer has considerable control over approach flight characteristics.

When entering final approach, an airplane is flying near point G of Fig. 8-1, and can be operating at speeds below point J during power-assisted STOL approaches. Note that if the airplane is allowed to go below speed N there will be no excess power available for altitude or attitude recovery. This is shown at point K, where no gap exists between the power-required and power-available curves; this means no power reserve remains for climb recovery if the approach is too low or too short.

Flaps assist here if they are the high-lift slotted or Fowler types previously reviewed. These flaps have the effect of reducing approach speed for a given rate of descent, or conversely, increasing the rate of descent for a given approach speed. Further, they provide considerably more control flexibility with power to maintain the required glidepath.

Such desirable flight characteristics are clearly shown by Fig. 8-2, which is based on actual flight data observed with a 150-hp Cherokee. With power and flap deflection stabilized for each curve, a direct reading was obtained for descent rate vs. airspeed, showing that for a fixed throttle setting and airplane configuration, airspeed was controlled by the stick. In other words, airspeed was controlled by the angle of attack (pitch trim attitude) of the airplane.

These curves also demonstrate the relationship between stick and throttle in controlling the airplane, particularly at low speed. While I do not wish to join

119

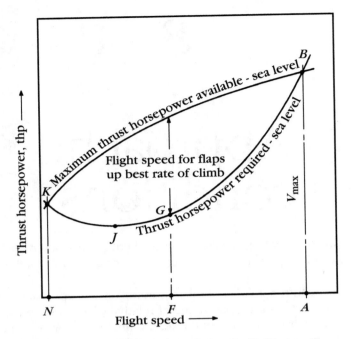

8-1 *Aircraft performance diagram illustrating low-speed power.*

the FAA-industrys flight instruction merry-go-round about the effect of stick vs. throttle on altitude and airspeed, it is quite obvious from Fig. 8-2 that both glide-path speed and slope corrections can be controlled through power changes. So the frequently expressed theorem that altitude is controlled by the stick should not be considered celestial law. In fact, I am sure most experienced pilots use both stick and throttle to apply crucial glidepath corrections, probably without realizing which one is applied at just what moment or in what sequence.

Note how the rate of descent can be varied along the approach path. In fact, descent rate decreases with airspeed right down to the stall—flaps up or down. This characteristic is common to all conventional aircraft and explains how a pilot tends to stretch a glide right into a full stall condition. Such accidents are most likely to occur in aircraft having small variations in control feel or trim attitude (angle of attack) with decreasing airspeed. There is an old theorem, "Make the pilot work hard to get into trouble," which means that stick forces should materially increase as the stall is approached. The Cardinal 177B does an excellent job of fulfilling this requirement, as is fully realized during the landing flare if elevator trim has been neglected. And this is the way the elevator control should be designed for safety.

To graphically depict the control options available, Fig. 8-2 shows that if you are a bit low in the approach and coming in power "off" with flaps up at 80 mph

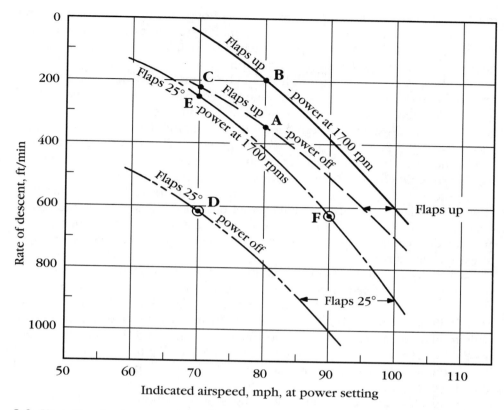

8-2 *Variation in descent rate with power and flap applications.*

at point A, adding a bit of power to 1700 rpm will decrease the rate of descent from 350 ft/min to 200 ft/min per point B. This is controlling altitude with the throttle. However, there is another alternative just mentioned—pulling back on the stick. Slowing from 80 to 70 mph, power "off" and flaps retracted, would reduce the rate of descent from 350 ft/min at A to 220 ft/min at C. Unfortunately, attempting to further reduce descent rate by pulling back some more will eventually terminate the approach in a spin.

The 25° flap position curves, with power "on" and power "off," show the same general relationship but because of flap drag have greater rates of descent for corresponding airspeed and power settings. For example, at point D the airplane is flying 70 mph at a power "off" descent rate of 620 ft/min with 25° flap extension. If things look marginally short, adding power to 1700 rpm will flatten the path to a descent rate of only 250 ft/min per point E. Some elevator adjustment is necessary to maintain 70 mph, but this can occur relatively slowly in modern aircraft; and possibly due to gust conditions, it might be considered desirable to approach a bit greater than 70 mph. Again, if the speed is reduced too much, a stall can occur with flaps extended, although the nose-high attitude and soft controls accompanying low-speed operation with flaps down should provide considerably more warning than when the flaps are up. This is another good reason for using flaps during final approach.

For these reasons, the approach should be controlled by throttle for descent rate and altitude, and by stick and throttle for the speed desired. For example, if gusty conditions exist in the landing area, it would be advisable to increase speed to 1.4 times stalling speed while remaining at the desired descent rate. As shown by Fig. 8-2 the 70 mph, 620 ft/min power "off" descent rate of point D can be increased by a safe 20 mph margin to 90 mph at point F by adding power to 1700 rpm. Of course, some elevator trim might also be required.

Figure 8-2 shows what can and does happen if the stick alone is used to control altitude and glidepath slope (rate of descent) until aircraft attitude takes over and a stall/spin condition terminates the flight. Remembering these speed and power relationships should improve everyone's approach technique, especially in rough air.

While discussing the use of flaps, I should mention pitch trim changes resulting from flap deflection, particularly since this effect could become quite crucial if not eliminated from the final design. To explain things simply, the effect of high lift flaps can be considered similar to increased lift acting along the wing trailing edge. As a result, flaps act to rotate the wing (and the entire airplane) nose-down about a spanwise pitch axis. In order to keep the airplane in a level flight attitude, this nose-down moment must be balanced by increased tail load and elevator forces.

Because of flap-induced airflow changes over the horizontal tail, some aircraft, such as the Piper Aztec and Beech Baron, tend to trim nose-up when flaps are lowered. So the type and amount of pitch change with flap deflection can vary from one airplane to another, making doubly important the need to design all aircraft for minimum trim changes with flap operation, both down and up.

If these unbalanced trim moments are large, the designer has three principal ways to reduce their effect on flight handling characteristics: (1) a responsive elevator trim system capable of handling the unbalanced moments at maximum flap deflection; (2) interconnecting the flap actuation and elevator trim systems to provide more nose-up trim as the flaps go down, and (3) limiting the amount of flap deflection.

The first method is the simplest and most effective, if it can be done. Regardless of the design approach used to control flap deflection moments, flight tests should demonstrate that control forces can be handled when the flaps are fully down and the trimming system disconnected.

Getting back to Fig. 8-1, when approach speeds drop below the minimum power-required point J, the power necessary to maintain level flight rises rapidly with decreasing airspeed. This seems contrary to the basic laws of physics because more power is apparently required to do less work. However, in the low-speed range between points J and K the airplane flight attitude (angle of attack) is such that total drag rises very rapidly and soon equals or exceeds the thrust available.

This region is known as the back side of the power curve, or the region of reversed command because more and more power is required to maintain control (command) at lower and lower airspeeds. It is a good flight area to avoid and generally no problem for our relatively low-speed private aircraft. However, you can find yourself there when practicing STOL approaches, particularly in one of the cleaner twin-engine aircraft.

Fortunately, there is an instrument that can offer increased safety in this region: the angle of attack (AOA) indicator. By displaying acceptable ranges of angle of attack for optimum cruise, as well as various flight conditions, the AOA indicator actually adds economy as well as safety to aircraft operation. Unfortunately, for some reason this flight instrument seems neglected by private aviation; I'm sure approach accidents would decrease in direct proportion to AOA indicator use and believe this device should be given more attention. If you are interested in pursuing this flight regime further, an article by Dan Manningham called "Flying Angle of Attack" includes material covering the subjects of stick vs. throttle, STOL operation, and flying angle of attack.[1]

Similar to the requirements for climbing and cruising flight, a positive level of static and dynamic stability about all three axes is important during the approach, particularly in IFR conditions. In addition, moderate control forces should be well balanced between all surfaces, adverse yaw should be very low, control response should be high right into the stall, and the longitudinal trim system should be effective.

While I have already discussed the design procedures required to obtain all these desirable qualities, it is interesting to note that FAA flight test approval requires approach speed demonstrations proving these characteristics are present at acceptable levels throughout the desired gross weight center-of-gravity range. One of these tests is a balked landing (missed approach) full-power climb-out without touching elevator trim. It is required to show whether a new design has undesirably high or reversing stick forces.

A second test is a tightening circular flight in approach trim, plus sudden application of full power. This test represents a negligent approach or an avoidance maneuver that must be accomplished without having the airplane fall out of the sky, or place extreme demands on the pilot. Of course, such tests place considerable demand on the airplane designer, and frequently they result in modifications to the control and trim systems along with control surface area, planform, and contour changes.

Responsive engine controls are essential during the descent and landing phases, probably to a greater degree than for any other nonacrobatic phase of flight. When power is needed, it is needed now, and not one-half second later due to a sloppy linkage or cable control. For similar reasons, effective carburetor heat control and response is also mandatory. Because of the number of accidents attributed to engine icing during the approach, possibly some consideration should be given to having carburetor heat linked to fully retarded throttle operation, as well as being independently controlled. This might eliminate another source of "pilot error" by automatically applying carb heat whenever the throttle is pulled all the way back.

All desirable approach characteristics are brought into sharp focus during IFR flight. Positive stability, minimum adverse yaw, effective surface controls with rapid response to very small deflections, a glidepath control system such as wing flaps, and positive engine controls all become very crucial in the fog—and separate quickly good, bad, and indifferent IFR designs. In addition, the ability of higher-performance aircraft to establish a steady, standard descent rate approximating 500 ft/min by lowering flaps, gear, or both when passing over the final approach fix is a source of great comfort and workload relief when conditions are piling up in the cockpit. The Piper Cherokee Arrow does this quite well with gear extension and without throttle adjustment at an approach speed of about 100 knots.

Because I thoroughly covered landing control and handling criteria during the crosswind discussion of chapter 5, it won't be considered again here; so our flight profile now closes with this brief approach review. In addition to providing some design background for the various segments of routine flight, I hope these chapters have improved your flight technique to some degree.

We next proceed to the important design and flight safety area of weight and balance control, followed by a performance summary section. Subsequent chapters include general and detail design information items that should be of interest and benefit to every pilot and aircraft owner.

Endnotes

[1](a) Dan Manningham, "Flying Angle of Attack," *Business and Commercial Aviation Magazine*, September 1975, p. 54.

(b) "Airplane Flight Characteristics and Techniques Associated with Operation in the Region of Reversed Command." *FAA Advisory Circular AC No. 61-50A*, March 15, 1976.

9

Useful load and weight control

Useful load, the last of our basic operational criteria, can make or break a new airplane intended for the serious business and pleasure markets. In view of this, every new design must provide as much flight performance and safety as possible at minimum empty weight.

Weight and balance terminology must be defined before covering the related calculations. *Useful load* is the difference between the empty weight established when the airplane was completed by the manufacturer, and the gross weight, which is the maximum legal flight weight. Because empty weight consists of the basic airframe, engine, instruments, and residual (undrainable) fuel and oil required to satisfy the minimum FAA airworthiness requirements, everything added until gross weight is reached is considered useful load. This means any primary blind or full IFR instrumentation, radios, antennas, wheel fairings, special interior trim, or special equipment (other than the mandatory emergency locator transmitter) is actually part of the fixed useful load. Such permanent additions rapidly reduce the disposable weight that can be carried in the form of oil, fuel, passengers, and baggage. Just a few options or modifications can have a crucial effect on duration and range, IFR suitability, and passenger-carrying ability—all basic parameters directly related to aircraft utility and acceptance.

You can gain a helpful background in this area from reviewing the design procedures followed in developing the maximum useful load for a new airplane.

Empty weight and preliminary design

The basic secret in designing the maximum useful load is getting and keeping a low empty weight for the powerplant size and design load factors required. And low empty weight means low airframe weight, which can only be realized through efficient structural analysis and system design.

For example, if the wing skin can be 0.032 in. thick according to stress analysis, the designer should not use 0.040 in. skin. In fact, if time and money permit, static test of a wing panel might prove that skin 0.025 in. thick is adequate. However, some trade-off may be required for this decision. If the airplane is intended for bush-country operation, it might be more prudent to leave the skin at 0.032 in. or go to 0.040 in. for a rugged structure, saving weight in some other area such as canvas sling seats instead of more comfortable upholstered ones. On the other hand, if the airplane is intended for the more gently operated private market, thin skin can be used and the weight saved converted to useful load or deluxe upholstered seating, both of which carry sales appeal.

Unfortunately, low structural weight usually requires thin skins and many small stiffeners and frames. This type of airframe design is obviously contrary to efficient, low-cost production and would result in an excessively expensive airplane. This approach may be acceptable for special experimental work to prove a new design concept, but certainly is not suitable for mass production as practiced by the private aircraft industry. So again the designer must compromise, this time balancing the empty weight goal with production labor-hour requirements. This takes time and partially explains why aircraft are expensive. Total engineering costs must be prorated over a predetermined number of production units, each carrying a portion of the design and development costs.

Somewhere around the first third of the total design period, someone always comes forward with a requirement for more horsepower; the theory apparently being that the final design can be made available for the same price if a larger engine is incorporated early enough in the design program. This is never so, since a larger engine and propeller cost and weigh more just sitting in their shipping containers; and they require heavier structure to support their own greater weight, as well as the increased flight speed they are supposed to deliver.

It is true, however, that if a new design is eventually to be developed into a line of similar aircraft having increased power and useful load, overall weight can be saved if the initial structure is designed with growth in mind. Although some penalty is incurred by the lower-powered, basic configuration, subsequent designs of higher power will be increasingly efficient. For a given gross weight and operational requirement, it is always less costly in both structural weight and final price to build strength into a basic structure rather than adding reinforcements and reworking the airframe at a later date.

Because close weight control is essential during the actual detail design phase, accurate weight estimates for the various airframe components must be developed during the preliminary design study program. These weights establish an empty weight target for each new airplane and should be carefully monitored right through certification.

Wing, fuselage or hull, tail surface, equipment, powerplant, and landing gear weights vary as a percentage of the total design weight, generally in accordance with operational use. As examples, the greater the applied load factor requirements, the heavier the structure; the more equipment necessary to perform a service, such as agricultural spraying, the larger the percentage of empty weight required for equipment items; and so forth.

Figure 9-1, prepared from detailed weights of many single-engine aircraft designed to recent FAR Part 23 requirements, is included as a preliminary design guide for basic weight estimation. It also indicates how structural weight increases with gross weight.

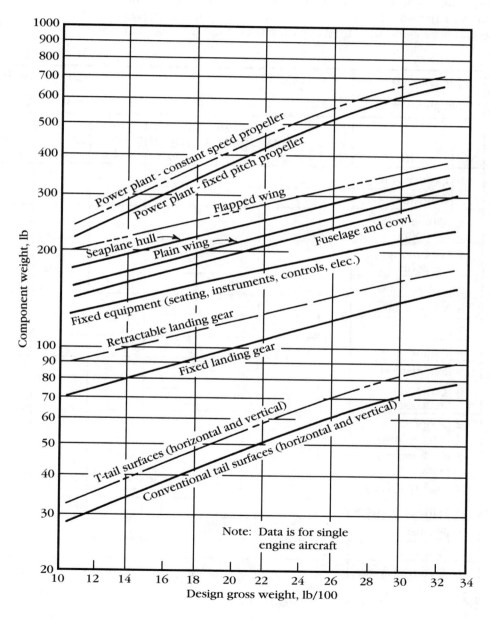

9-1 *FAR Part 23 aircraft component weight vs. design gross weight.*

Once these various weights have been more or less established, it is necessary to properly arrange the different airframe and useful load items so the airplane will balance throughout all flight and loading conditions. These preliminary weight and balance calculations are usually presented in a form similar to Fig. 9-2. This is known as the preliminary weight and balance summary. It is prepared in the same way as the individual calculations and log entries necessary any time equipment items are added to or removed from a licensed airplane.

Page _____

WEIGHT AND BALANCE SUMMARY							L.E. MAC @ Sta. _____		
Configuration _____							MAC = _____ in.		

No.	Item	Wt, lbs	Horizontal arm, in.	Horizontal moment in. lbs	Vertical arm, in.	Vertical moment, in.-lbs	Alternate wt, lbs	Arm, in.	Moment, in.-lbs
1	Fuselage/hull								
2	Windshield								
3	Sliding canopy or doors								
4	Aft canopy or doors								
5	Horizontal tail								
6	Vertical tail								
7	Wing								
8	Engine								
9	Engine mount								
10	Cowl and exhaust system								
12	Propeller, gov. and spinner								
13	Carburetor air box								
14	Fuel & oil system								
15	Surface controls								
16	Furnishings								
17	Instruments								
18	Electrical system								

9-2 *Weight and balance summary form.*

19	Main gear (retractable)										
20	Nosewheel (retractable)										
21	Hydraulic system										
22	Battery inst.										
23	Reserve fuel and oil										
Empty weight											
24	Pilot and passengers										
25	Rear passengers										
26	Oil (_____ qt)										
27	Fuel (_____ gal)										
28	Accessories (radio, etc.)										
29	Baggage										
Useful load											
Gross weight											

$$\left(\frac{\text{Horizontal arm—L.E. MAC Sta.}}{\text{MAC}}\right) = \begin{array}{l}\text{c.g. location from MAC leading edge in fraction}\\\text{of MAC (multiplying by 100 gives percent MAC)}\end{array}$$

Model ————————————————   Report no. ——————————

Cont ————————————————   Date ————————————

9-2 *Continued.*

Every designer is faced with the problem of providing center-of-gravity limits that will accept all or most loading conditions necessary for convenient and useful operation. Since flight tests must demonstrate the existence of adequate stability at the extremes of the center-of-gravity envelope, the desired center-of-gravity range may influence horizontal and vertical tail surface design. Because longitudinal, lateral, and directional stability are all affected by tail surface area, the need for a center-of-gravity range extending far aft may result in increased tail arm or areas, or both, as discussed in chapter 7.

A desirable center-of-gravity range for powered aircraft extends from about 15 to 28% of the average wing chord, or more correctly referred to as the *mean aerodynamic chord,* or MAC. Sailplane designers prefer the center of gravity further aft to keep the horizontal tail relatively unloaded, and so reduce tail area and drag. Such a center-of-gravity location is usually accompanied by neutral or slightly positive longitudinal stability and very low stick forces, which are not acceptable for private aircraft operation.

As shown in Fig. 9-3, to obtain the fore and aft location of the center of gravity, the weight of each component is multiplied by its distance (or arm) measured from a vertical reference line called the *horizontal datum*, to the center of gravity of the component. These horizontal moments are summarized in Fig. 9-2. To obtain the horizontal center-of-gravity position, the total of the horizontal moments is divided by the total weight of all the items used to compute the total horizontal moment, in this manner

$$\frac{\text{Summation of horizontal moments}}{\text{total weight of items}} = \begin{array}{l}\text{c.g. location from the vertical reference}\\ \text{(or datum) line in inches}\end{array}$$

9-3 *Horizontal and vertical datum line locations, with an example shown at the center of gravity for the pilot.*

The vertical center-of-gravity position is determined by the same procedure except the arm to each weight item is measured from a horizontal base line, or *vertical datum*, and then summarized on the form of Fig. 9-2 alongside the horizontal moment values. The reference datum lines for these fairly simple, but laborious calculations can be conveniently located, as shown in Fig. 9-3; they will be clearly marked on the airplane side view included with every aircraft weight and balance statement.

The center-of-gravity location as a percentage of the MAC is simply determined as follows. Again referring to Fig. 9-3, if the MAC leading edge is at station 110 and the MAC is 60 in., when the center of gravity is at station 125 we have c.g. location referred to the MAC.

$$\text{c.g. location referred to the MAC} = \frac{(\text{c.g. location}) - (\text{leading edge MAC station})}{\text{MAC}}$$

$$= \frac{125 - 110}{60}$$

$$= \frac{15}{60}$$

$$= .25 \text{ (or 25\% MAC when multiplied by 100)}$$

The form for this calculation is shown at the bottom of Fig. 9-2. After the balance range has been computed within acceptable limits, the center-of-gravity position should be closely monitored during the entire design period. A high level of weight supervision is mandatory to preserve center-of-gravity limits and avoid added ballast or radical last minute changes in equipment location. Unfortunately, weight control is all too frequently insufficient to provide the direction required; and so some fast changes usually occur immediately after the first weighing of a new design.

Active weight control frequently results in redesign of a structural component that is becoming too heavy. Although removal of excess weight may be costly in time, engineering, and materials, it is essential if the desired useful load is to be achieved.

Current FAA requirements add to the empty weight all the fuel and oil that cannot be readily drained on the ground or used in flight, whichever is greater. Modern aircraft have wing fuel tanks, and the amount of this residual fuel is determined during a steady slip or skid, again whichever maneuver results in more fuel remaining in the tanks when the fuel pressure begins to drop off.

Because wing tanks are flat by nature, a considerable volume of fuel can be piled up away from the tank outlet. Most of this fuel could be used in normal, level flight, but the FAA considers as crucial the possibility that someone will run dry during a steep slip or skid on final. This could happen if fuel is low or the tank on the slip side is in use, both of which indicate poor fuel management but occur regularly.

Fortunately, it is possible to contain sufficient fuel at the tank outlet during maneuvers to minimize the amount of unusable fuel in wing tanks, as covered in chapter 11 under fuel systems. Unfortunately, all residual fuel weight established by these rather extreme tests must be included in the certified weight statement as part of the empty weight, and, of course, the useful load is reduced accordingly. On some small aircraft this value approaches 30 lbs, representing the weight of a good dual nav/com radio. As a result, increased attention needs to be paid to the detailed design of fuel systems, which is covered in chapter 11.

Useful load

The weight and balance statement supplied with each airplane is obtained by actually weighing the completed airplane prior to delivery. These statements have legal value since they represent the licensed weight empty and are part of the FAA Certificate of Airworthiness requirements applying to every production airplane sold. This weight statement forms part of the airplane flight manual carried in new aircraft, or is part of the owner's manual supplied with older airplanes. As such, the weight statement must remain in the airplane at all times.

As discussed earlier, the difference between the actual certified weight and the design gross weight represents the total allowable useful load. However, a good percentage of this value might have been permanently removed through installation of IFR instruments and avionics, special equipment additions, and structural repairs or modifications. What is left is the disposable load capacity available to carry oil, fuel, passengers, and baggage.

To ensure allowable gross weight or approved center-of-gravity limits are not exceeded, every pilot must be fully aware of the useful load being carried on each flight, as well as the way this load is distributed in the airplane. Just because an airplane is advertised as having a 1000-lb useful load, do not assume 1000 lbs can be added to that ship parked on the ramp. The Certified Weight and Balance Statement as amended to include current changes in equipment and airframe modifications is the sole source and authority for useful load capability.

At the risk of being accused of beating a dead horse, the danger of overloading must be mentioned again since private aircraft performance deteriorates rapidly with overload even under standard air conditions; to say nothing of performance losses under high ambient temperatures or pressure altitudes as covered in chapter 6. A set of typical flight manual load and moment diagrams are shown in Figs. 9-4a and 9-4b.

Since stalling speed varies directly with the square root of the weight ratio

$$V_2 = V_1 \sqrt{\frac{W_2}{W_1}}$$

for each 10% increase in gross weight, the stalling speed increases by 5%. This means more than a 5% longer takeoff run assuming comparable ambient temperature and pressure conditions, followed by a decreased ROC.

As indicated by the climb formula in chapter 6, climb rate is inversely proportional to aircraft weight, assuming no variation in climb drag with aircraft weight. This is shown by the ratio

$$ROC_2 = ROC_1 \left(\frac{W_1}{W_2} \right)$$

which roughly indicates a 10% reduction in climb for a 10% increase in weight. However, the airplane drag coefficient will increase slightly with weight since the wing assumes a greater angle of attack to carry the heavier weight. This in turn increases the induced drag, as well as the overall drag resulting from increased fuselage angle of attack and interference drag. So our ROC actually decreases faster than the simple weight ratio indicates, and often catastrophically so.

Accident data published by the National Transportation Safety Board in 1974 list 40 to 70 fatalities per year directly attributable to poor weight and balance management.[1] In a number of crashes, Cherokee 140s or Cessna 150s were being flown greater than 240 lbs over gross, and in at least one instance the center of gravity was 2 in. aft of the approved rearward limit. The latest NTSB weight and balance accident briefs, covering the period from January 1990 through December 1992, include 75 such accidents occurring at an average rate of 25 aircraft per year.[2]

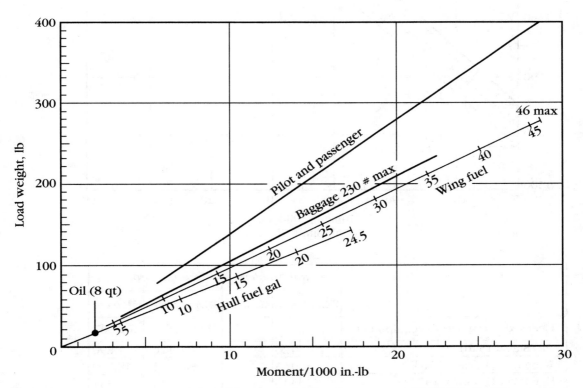

9-4a *Chart of moment values for different useful load items. Procedure: (1) Find the point on the line corresponding to the flight loading and record the weight and moment in the table. (2) Locate the final weight and moment totals (example for 2112 pounds at 219,311 in.-lb) on the weight-moment chart of Fig. 9-4b. The final weight and moment values must fall within the center of gravity limits established by the lines indicating forward or aft center of gravity.*

Moment/1000, in.-lb
weight limit: 2200 lb

	Actual problem			Sample problem		
	Weight	Arm	Moment	Weight	Arm	Moment
Licensed empty weight/equipment		—		1454	—	162,867
Oil (8 qt = 15 lb)		126		15	126	1,890
Wing fuel (46 gal = 276 lb)		103		180	103	18,540
Hull fuel (25.5 gal = 147 lb)		118		3	118	354
Pilot and passenger		71		340	71	24,140
Baggage		96		120	96	11,520
Totals				2112	103.8	219,311

9-4b *Chart locating total load and moment values vs. center of gravity limits. Flight must be conducted within the center of gravity limits shown at any weight.*

Unfortunately, fatalities remain comparatively high—totaling 54 for 1990, 25 for 1991, and 42 for 1992 related to improper and/or overweight loading. So weight and balance fatalities continue to remain too high.

Please take care. There is only so much a designer can do to make flying safe and enjoyable; and overloading has never been an approved means of increasing useful load.

Weight and balance control records

The aircraft weight and balance statement supplied by the manufacturer must be modified each time new items of fixed equipment or structure are added to the airplane, as well as whenever any previously installed equipment or structure is permanently removed. This is a legal requirement of FAA flight regulations intended to ensure that persons using an airplane will have the current weight and balance values at their fingertips along with the aircraft flight manual.

The owner of every airplane is held responsible for maintaining current weight and balance control records, regardless of whether the owner is an individual, a fixed-base operator, or any type of corporation. Maintaining these records is a simple procedure and one that is frequently practiced by every FAA-licensed aircraft mechanic. However, you may be able to save some time and money by working out your own weight statements whenever any changes are necessary.

The following brief example provides the background necessary to check the mechanic's calculations the next time some modifications are required. The basic calculations for adding a second set of radios plus another communications antenna may be tabulated in this manner:

	Weight	**Horizontal arm**	**Moment**
1. Aircraft weight empty plus fixed equipment (per factory weight & balance)	1435#	86.06"	123,500"#
2. Add second nav/com	15	65.50	983
3. Second com antenna	3	120.75	362
4. Flight line weight & c.g.	1453#	85.92"	124,845"#

The values across line 1 are taken right from the current weight and balance statement. These numbers are those supplied by the factory for an airplane not previously modified, or the latest weight log totals for a ship that has some changes. At any rate, to these basic totals is added the weight and horizontal location of the new nav/com as shown in line 2. This location is the horizontal distance, or horizontal arm, from the aircraft horizontal datum to the center of gravity of the installed avionics equipment per Fig. 9-5.

9-5 *Location of items for the weight and balance calculations required by the addition of a second nav/com radio and antenna.*

The equipment manufacturer shows this center-of-gravity position in the installation data, and presumably you or your mechanic can measure from some known airframe reference point to the equipment center of gravity. Usually the aircraft flight manual or owner's manual will list some structural reference station such as a door frame or wing leading edge for this purpose.

Multiplying the nav/com weight by this arm provides the moment for the installed radio, 983 in.-lb for our example. Likewise, the second antenna is added into the weight and balance calculation as shown by line 3. Note that fractions of in.-lb have not been included when they are less than 0.50; if 0.50 or larger the next full in.-lb is used. For our installation 982.5 becomes 983 in.-lb and 362.25 is listed as 362 in.-lb.

We next add the weights on line 4 and find the new total of 1453 lbs. This is properly referred to as the flight line weight since this value represents the equipped empty weight of the airplane before the disposable useful load has been added. Note, however, that there is now 18 lbs less capacity available for passengers, baggage, fuel, and oil as a result of the decrease in disposable useful load due to the added fixed weight of the avionics equipment.

This equipped empty weight is frequently referred to as the airplane empty weight, which is acceptable practice as long as no one believes it is the same as the basic airplane empty weight before fixed equipment has been added. This would mean the full useful load could be carried, and such a misunderstanding could be the first step toward an overload accident. As a matter of good practice, inspect the airplane weight and balance log whenever you first fly any airplane; and make sure you have the latest weight statement. Frequently the weight changes go on for many pages with no reference to the current final statement—a bad practice, but a very common one.

As a final step, adding up the new moment total and dividing by the new weight of 1453 lbs gives us the new horizontal arm of 85.92 in. from the datum to the flight line weight center-of-gravity position. Note that the airplane center of gravity has moved forward a bit from the original 86.06-in. location, as would be expected when equipment is added ahead of the instrument panel.

Regardless of the type of weight change, the preceding procedure will supply the required answer. If you are interested in this subject, the endnotes carry more complex weight and balance examples of varied detail.[1,3] The *FAA Advisory Circular* is also of particular interest if structural modifications or repairs must be documented in accordance with FAA requirements.[3]

I believe one of the most important facts an aircraft owner can remember is that from the time a well-designed airplane first flies, everything done to it contributes to increased empty weight, reduced performance, and greater cost. This holds true throughout the life of the airplane; like cruising boats and human beings, aircraft seem to grow heavier with age. And since useful load, as well as takeoff and climb performance, decrease with increased weight, both utility and safety suffer from every addition of unnecessary fixed weight.

Endnotes

[1]"Weight and Balance: An Important Safety Consideration for the General Aviation Pilot." *National Transportation Safety Board Report NTSB-PAM-74-1*, December 18, 1974.

[2]"NTSB Briefs of Accidents Involving Weight and Balance as Cause/Factor/Finding." *National Transportation Safety Board Report NTSB/ARG-93/01*, January 1990–December 1992.

[3]Pages 243–265 of *FAA Advisory Circular AC 43.13-1A*, "Acceptable Methods, Techniques, and Practices—Aircraft Inspection and Repair." Available from: Superintendent of Documents, U.S. Government Printing Office, 710 North Captiol St., N.W., Washington, DC 20402, Order Publication FAA AC No. 43.13-1A, stock No. 5011-0058.

10

Operating performance

In a sense, this chapter is a summary of all preceding design data; by weighting the various parameters for each phase of flight it is possible to show their individual importance for aircraft design. Since this information is of interest to pilots as well as those concerned with aircraft design, it is presented before beginning a more detailed performance review. Acronyms and symbols are defined at the end of this chapter.[1]

Weighting design parameters for flight conditions

I have divided aircraft parameters into 11 categories: ground handling, power, constant-speed propeller, wing area, aspect ratio, flaps, low-speed control, stability, drag, responsive engine controls, and weight. The more important the contribution to safe and acceptable flight characteristics for the specific operation, the higher the unit value, with 10 being the maximum number assigned for any item.

Admittedly, the priority listings can appear arbitrary to some, in which case the values can be reassigned and a new set of priorities developed. However, the parameters will probably end up in a quite similar final order. From a design standpoint, the relative values used in Table 10-1 apply to current production aircraft and so represent a working demonstration of modern private aircraft design practice.

Although all of these parameters have some obvious value and importance to aircraft design. To obtain a listing of relative design merit, add the priority values for each one in each phase of flight. When arranged in the order of decreasing totals, and therefore importance to modern aircraft design, the following results:

1 Stable airplane—37 points
2 Horsepower—36
3 Constant-speed propeller—36
4 Low weight—35
5 Low-speed control—29

6 Large wing area (comparatively low-wing loading)—26
7 Clean airplane (low-drag design)—25
8 Flaps—23
9 Ease of ground handling—20
10 High aspect ratio—18
11 Responsive engine controls—14

Small wing area, which is necessary for very high-speed cruise, represents a special condition not relevant to normal private aircraft operation, and so finishes well down in the rating with only 6 points. However, if speed were paramount, small wing area would be weighted differently and probably would replace large wing area at the number 6 position. The point being that the data in Table 10-1 was prepared for a typical FAR Part 23 private aircraft design, and should be considered primarily for that class of airplane.

A stable airplane configuration is of primary importance, as would be expected under current FAA design requirements and when you consider flight safety. This parameter designs the tail, wing, and control surfaces. Previous chapters might have pointed out the cost and weight penalties associated with horsepower. The correct selection and use of horsepower, or more properly thrust horsepower, is important to every basic airplane design because it applies in some manner to all phases of powered aircraft flight. In view of this, it is not surprising to find power-plant parameters tied for second- and third-place importance in our evaluation.

Low weight appears high on the list because it is not only an indication of design and structural efficiency but also is important throughout the entire life of the airplane. For a given task, the lighter airplane should cost less and be more economical to operate.

Comparatively large wing area, low-speed control, and flaps all apply to safe low-speed operation from small airports and under crosswind conditions. These criteria are grouped to form a middle of merit rating that binds together the primary and secondary criteria, items 1 through 4 and 9 through 11 respectively.

It is surprising to find low drag in seventh position. The reason is, although low drag is desirable throughout the entire flight spectrum, it is only crucial for three of the six flight conditions listed. Because of the minimum effect of drag at low speeds, this parameter does not seriously enter takeoff or landing considerations for private aircraft design, while a high drag value might actually be desirable during the approach. And of the three conditions affected by low drag, only climb and cruise are important; service ceiling is rarely called for in our type of flying, although it deserves more consideration for mountainous areas of the West, Canada, South America, and Africa.

High aspect ratio reduces induced drag during climb and cruise, releasing additional power to increase climb rate and cruising speed. However, in return for this gain is a heavier wing structure, which increases the empty weight and the gross weight if a constant useful load value is maintained. Because climb varies inversely with weight, some of the climb provided by increased aspect ratio is lost to increased weight.

Table 10-1 Design parameters for flight conditions, FAR Part 23 airplane, weighted by importance to each phase of flight operation

Takeoff	Climb	Cruise	Service ceiling	Descent and approach	Landing
10. Ease of ground handling and control (tricycle gear)	10. Thrust hp margin (constant speed propeller for maximum thrust)	10. Stable airplane	10. Maximum hp margin (turbo-charged engine for maximum alt. perf.)	10. Stable airplane with minimum adverse yaw	10. Tricycle landing gear
9. Thrust (hp and constant speed propeller for maximum thrust)	9. Low weight	9. Clean airplane	9. Constant speed propeller to deliver maximum thrust	9. Flaps for speed and descent control (or use of spoilers, lowering landing gear, etc., to control drag value)	9. Aileron, rudder and elevator control systems that are responsive at low speed with minimum adverse yaw
8. Low speed rudder and aileron control (elevator also for all seaplanes and STOL landplanes)	8. Clean airplane to reduce climb drag (faired landing gear, clean cowl, wing root fillets, etc.)	8. Constant speed propeller to provide takeoff and climb without penalizing cruise performance	8. Clean airplane to reduce drag	8. Responsive surface and engine control systems	8. Combination of wing area and flaps adequate to provide low landing speed (particularly important for seaplanes and short-field operations)
7. Low weight	7. Large wing area and high aspect ratio to reduce C_L and induced drag: $$C_{D_i} = \frac{C_L^2}{\pi AR}$$ C_L varies inversely with S_w and therefore C_{D_i} decreases as the square of lift reduction—also C_{D_i} decreases with AR increase	7. Relatively high power to permit lower rpm and quieter cruise	7. Low weight	7. For IFR work, the ability to rapidly set up a 500 fpm rate of descent over the approach fix	7. Low weight for low landing speed
6. Wing area and flaps to provide low takeoff speed (particularly important for sea- and short-field operaton)	6. Stable airplane	6. Small wing area and high aspect ratio to reduce parasite and induced drags (a real trade-off required here)	6. Stable airplane (cannot attain maximum ceiling if control surfaces must be constantly operated to maintain desired flight path)		6. Responsive engine and throttle system for control in gusty air or aborted landing
		5. Low weight	5. Large wing area and high AR to reduce C_L and induced drag		5. Stable airplane
			4. Responsive surface controls with minimum adverse yaw		

Ease of ground handling and responsive engine controls are self-evident, although I'm sure that many student pilots and their instructors frequently consider ground handling of primary importance. Again as a reminder, the merit order of these design parameters is based on requirements for a private airplane design and is not weighted toward trainer use. However, you can easily rearrange the priorities for such an application yourself to determine how the resulting airplane should be designed. Stability would still be paramount, but power would drop in importance while ground handling and low-speed control advance in rating. The result: a Cessna 152, Piper Cadet, or perhaps the old Piper Cub and Aeronca Champ.

What price performance?

Competitive aircraft performance is just as important as efficient structural design and handling characteristics in determining acceptance. If an airplane does not climb or cruise at least on a level with others of its class, it will not be successful regardless of how structurally sound it might be or how easily maneuvered. Since the performance diagram ranks in equal importance with the V-n envelope (Fig. 2-2) in detailing aircraft characteristics, I'll briefly review the information carried by the performance curves of Fig. 10-1. Performance formulas are given at the end of this chapter.[2]

At speed A, the power-available and power-required curves intersect at B, the maximum speed possible at the altitude for which the curves have been derived. No more power is available, so the airplane cannot exceed power-required speed A. When the engine output is reduced to 65% power, the intersection at D indicates cruising speed C is obtained at this power setting. The thrust horsepower available between D and E represents the power available to climb at speed C when the airplane is trimmed for 65% power cruising flight at point D.

These power-available and power-required curves are usually prepared for sea level standard conditions. However, for performance comparison, power-required values might be given for other altitudes similar to the 7000-ft curve through point D', while additional power-available curves would be prepared for supercharged engine operation.

Although considerable calculation is required to obtain power required vs. speed, this curve is sometimes presented for more than one weight or design configuration if such data are necessary for important operating conditions (such as switching from faired landing gear to seaplane floats, which affect both weight and drag). Speed conversion values are given at the end of this chapter.[3]

Although detailed performance calculations are quite lengthy and so are not covered in this book, endnotes 4–10 contain a list of suitably comprehensive sources for those interested in developing performance diagrams.

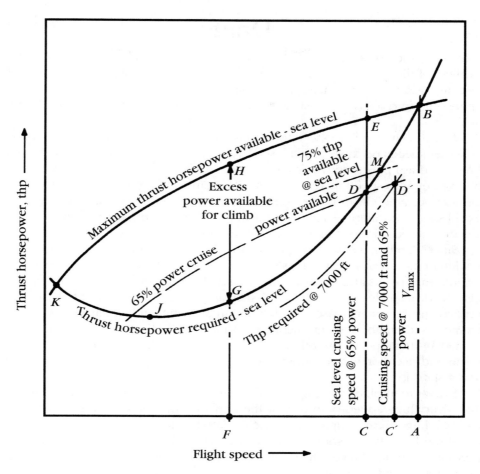

10-1 *Airplane performance diagram.*

The various design features affecting the climb, cruise, and approach phases of the performance diagram have been thoroughly covered in chapters 6, 7, and 8 and will not be mentioned again. As a summary, Table 10-1 shows the effect of various options on aircraft performance. High speed requires high power, low wing area, and low drag; climb is improved by adding high aspect ratio wing area, increasing power, reducing weight, or by all three; and cruising speed is related to moderate wing area, an efficient powerplant, and low drag.

Because drag is mentioned with every reference to efficient operation and has not previously been covered in depth, this is a good point to review drag causes and cures, particularly because drag is basic to all performance evaluation, comparison, and improvement.

Drag

Like death and taxes, drag is always with us, and it is just about as welcome. Regardless of flight attitude or speed, drag forces are continually exerting some effect on aircraft performance and handling characteristics. In fact it is virtually impossible to discuss drag without also considering performance and handling qualities.

As we realize only too well today, automobiles consume more fuel per mile at higher speeds because of increased bearing, drive system, and engine friction (internal drag) plus aerodynamic resistance (external drag), both of which must be overcome as speed increases. In similar manner, displacement boats require tremendous amounts of power to increase speed slightly beyond a design speed established by waterline length. And depending on hull shape, might not be able to exceed this speed regardless of the power available.

Since aircraft have three degrees of freedom of motion, drag affects their operation in more ways than it does surface vehicles. In addition to lowering cruising and top speeds and thereby increasing fuel consumption, drag, as previously discussed, also decreases ROC and increases the rate of descent during approach; this in turn increases the skill required during landing flareout. Although an increased rate of descent might be desirable on final approach when accompanied by low speed, the other negative characteristics certainly are not welcome and occasionally become the deciding factor between a successful flight and a statistic.

Similar to airplane weight, drag also has an insidious way of increasing with age. Everything added onto or into the airplane contributes just a bit more additional drag. Because drag is so closely related to aircraft operation, consider in detail the basic types, sources, and causes of airplane drag. In addition, some performance improvement features will be noted; these should benefit your airplane and would be incorporated into any new design.

What causes drag?

Airplane drag is generally considered the sum of two different types: parasite drag and induced drag. From the designer's standpoint, it is preferable to separate drag into three components: parasite drag, wing profile drag (from airfoil section drag), and induced drag. Isolating wing profile drag from parasite drag gives the designer a direct indication of the effect of various airfoil sections on cruising flight. This is an important consideration because the lower the wing profile drag at cruise lift coefficients, the greater the speed range from stall to maximum level flight speed, and the more efficient the airplane.

Just what causes these types of drag? Their source falls into eight general categories. The total value of drag types 1 through 7 represents the airplane parasite drag and this sum plus the induced drag is the airplane total drag. The various total drag components are:

1. Flat plate drag

Due to the frontal area of the airplane. The cross-sectional area of the fuselage (or hull) and other structural members such as exposed landing gear, floats, or skis can be considered as parasite drag and explains why high-performance aircraft are designed to have minimum frontal area and retractable landing gear.

2. Skin friction drag

Air passing over a surface creates drag in proportion to the amount of exposed area and the roughness of the surface. For this reason, high-performance aircraft are designed with flush-riveted fuselages, tail surfaces, and wings (which are also of minimum surface area), while high-performance sailplanes are designed with minimum total surface area which is then hand-rubbed to a polished finish.

3. Interference drag

Caused by the breakdown of airflow over external structural and accessory items, such as landing gear struts, wheels and skis, wing and tail surface brace struts, flying wires, exposed bolt heads and nuts, drain fittings, radio antennas, and similar appendages. Frequently, this turbulent air flows over other parts of the airplane causing additional drag plus a possible reduction in surface effectiveness, control response, and aircraft stability.

4. Separation drag

Occurs from the overexpansion of air flowing over rough or poorly designed intersections and structural contours, such as abrupt changes in fuselage shape, the sharp break of hull lines at the transverse step in a hull bottom, irregular contour of the wing leading edge and upper surface, the need for wing root area fillets, cowling flow breakdown, and protruding canopy shapes—to name a few causes. Separation drag might frequently be of an alternating attaching/detaching flow type. This problem is particularly common with amphibians and can manifest itself as a most annoying cause of left/right hunting in yawed flight, with the airplane first heading off about 10° in one direction and then swinging back to a similar displacement in the opposite direction. The source of such a flow breakdown can be difficult to isolate and equally challenging to cure, but it is usually associated with flow from a cowling, wing root, landing gear, or strut intersection that changes the pressure field flowing over the vertical tail, which in turn causes the oscillating flight.

5. Pressure field drag

Somewhat similar to a combination of interference and separation drags. When leakage flow occurs around or past poorly fitting doors, removable panels, rudders, elevators, ailerons, and flaps, the change in pressure due to flow separation or induced turbulence creates additional drag. This condition is probably most apparent as whistling doors (due to poor sealing) and leaking cowling panels; but flow leakage around control surfaces is probably a greater source of drag at higher speeds, and reduces control response at all speeds.

6. Powerplant drag

Represents the energy used to cool the cylinders and oil system, provide combustion air, and drive accessory items such as fuel, hydraulic, and vacuum pumps, prop governor, and alternator. The amount of cooling air required during climb usually designs the air inlet and exit areas. For higher-powered aircraft, cooling flaps are frequently used to provide adequate outlet area during climb, and are then closed for minimum cooling drag during cruise flight.

7. Wing-section profile drag

Varies with the airfoil section selected by the designer. For maximum cruising range, and for high-performance aircraft, a low value of wing-section profile drag is desirable. The airfoil should be selected with low profile drag, low section moment, and gentle stall characteristics as primary considerations. As noted, by keeping the airfoil-section profile drag separate from the airplane total drag, the designer can readily determine the effect of different airfoils on the operational characteristics of a new design. (See the endnotes in chapter 6[2] for airfoil data.)

8. Induced drag

This category of airplane drag is particularly unique because it is composed of two complementary parameters that can also make or break a design from the moment the configuration is established. These parameters are wing area and aspect ratio, and both affect the magnitude of induced drag. Since induced drag can represent one-half to two-thirds of the airplane's total drag value at climb speeds, I will again discuss the effect of wing area and aspect ratio.

At a given weight and speed, the lift coefficient necessary to sustain level flight varies with wing area: the more wing area, the lower the required value of lift coefficient C_L, as shown by

$$C_L = \frac{\text{aircraft weight}}{\text{air density} \times \text{wing area} \times (\text{velocity})^2}$$

And the lower the C_L value the lower the induced drag coefficient C_{Di} because induced drag is determined by the equation

$$C_{D_i} = \frac{(C_L)^2}{\pi(AR)}$$

where C_{D_i} = induced drag coefficient
C_L = lift coefficient
AR = wing aspect ratio

So you see that aircraft intended for moderately slow flight, which includes most of our 115 to 150 mph general aviation types, have undesirable built-in induced drag due to high C_L values unless designed with adequate wing area. While increasing wing area also reduces landing speed and improves climb performance, the wing weight, frontal area, and skin friction drag will be greater; so a design compromise is necessary. Chapter 6 covers the effects of aspect ratio and induced drag in great detail under the subject of increasing climb. Refer to that chapter if additional information is desired.

Reducing drag

I know that many of you are not concerned about drag reduction if annual flying time is quite low. However, when flying as little as 50 hrs a year, a 10% reduction in drag will effectively provide a few hours of free flying. For example, assuming flight over the same distance, the savings possible from a 10% drag reduction for an airplane using 9 gal/hr of $2/gal fuel is:

```
Original annual cost = 50 hr × 9 gal/hr × $2          = $900
10% drag reduction  = 50 hr × (0.9 × 9 gal/hr) × $2  = $810
                          Annual savings in fuel alone = $  90
```

Everyone would enjoy flying faster for the same price, or being able to reduce power and fly at the former cruising speed with less noise and cost. And as added gains, the ROC increases with a cleaner airplane, while higher cruising speed with the same fuel consumption is an obvious safety factor when skirting weather or operating in IFR conditions.

There are a few simple ways to accomplish drag reduction on any airplane. In fact, some are associated with routine maintenance. Although it is apparent that little can be done to reduce drag types 1, 7, and 8 because they are built into the design, there is ample opportunity for performance improvement by working on 2 through 6. However, even type 1 flat plate drag can be partially overcome when selecting a new airplane; that sleek design with the somewhat tighter cabin and thinner wings will go faster for the same horsepower. This is the secret of the successful line of Mooney aircraft.

Keeping an airplane clean and polished is a matter of performance efficiency as well as proper maintenance. The smoother the surface, the lower the total skin friction drag. This advice also applies to keeping wing and tail surface leading edges free of mud, insects, and dents; they disturb the airflow, adding to drag and reducing maximum lift. While the outer panels can't retract in flight to reduce wing area, the entire airplane—or at least the wing and tail surfaces—can be kept clean.

The elimination of interference drag probably offers the maximum opportunity for performance improvement. In landing gear design alone, comparison studies noted in chapter 6 indicate that adding wheel fairings to tricycle gear aircraft will improve climb performance midway to that of a fully retracted gear.

Taking some pains to fair over or enclose exposed nuts, bolts, drain fittings, and similar small items will not only reduce interference drag but also eliminate some likely sources of early icing. And those strut fairings that no longer fit tightly, are misaligned, or possibly just missing, are another fine source of drag and ice buildup. They should be tightly in place and positioned to align with the airflow direction.

Although the various antennas carried today might not lend themselves to arbitrary placement, particularly if the equipment they serve is to work properly, it is usually possible to locate them where turbulent air flowing off an antenna will not fall directly on a strut or landing gear attachment fitting. These structural joints are probably producing turbulent flow all by themselves without any additional help from the antenna.

For most applications, a small whip-type communications antenna will serve very well and with much less drag than a large blade, while the short ball-ended transponder antenna also works well with DME (distance-measuring equipment) equipment and at less drag and cost than the usual blade antenna (as discussed in chapter 13).

The climb and cruise performance of many aircraft can be improved by adding wing root and canopy fairings to reduce flow separation. Although such additions are not for the amateur, proper filleting can reduce drag and improve stall characteristics.

If you suspect flow breakdown in some area of your airplane, tuft the region using pieces of dark yarn about 4 in long. They can be attached at their forward end with masking tape and can be observed in flight from the cockpit of another airplane flying 100 to 150 ft away. If the pieces of yarn oscillate a half-inch or more or if they reverse to curl forward in flight, some degree of flow breakdown is indicated.

If you want to improve the flow, have a qualified aeronautical engineer recommend appropriate design modifications and approve any flight tests. Professional advice is suggested because correction of flow separation can occasionally have a pronounced effect on flight stability and handling characteristics.

Reducing pressure field drag becomes increasingly important as an airplane gets older, presenting worthwhile possibilities for drag improvement by eliminating various leaks and openings. Doors should be carefully reshaped to fit the fuselage contour, and additional rubber seal strip added if necessary. This effort reduces both drag and cabin noise level. Small access holes are frequently drilled in skins to accept socket wrenches or permit visual inspection; these can be carefully taped over or plugged with a flush plastic cap that can be removed as required for servicing.

Loose cowling panels should be refitted and attachment hardware adjusted or replaced to assure positive, nonleaking contact. The cooling air should enter at the cowl front and exhaust through the designed outlet area, not all along the cowling sides, top, and bottom.

If you have the time and interest, it is possible to continue the list of improvements to include more efficient cooling airflow, sealed control surfaces, etc. Since many of these possibilities require fairly sophisticated engineering support, a basic drag reduction program might be started by simply cleaning and polishing the entire ship, adding wheel fairings, and getting the doors and cowling tight. Performance should be noticeably improved by such attention. With today's fuel prices, any time spent to reduce drag will be amply rewarded, and you will have a better and safer airplane.

Drag can frequently be reduced by simply getting the airplane trimmed for maximum cruising speed at a given power setting. Sometimes referred to as "getting the airplane onto the step," a worthwhile increase in cruise will be realized for most aircraft by first entering a shallow dive and then carefully trimming in level flight to maintain as much of the speed increase as possible. The cruising speed of hull-type amphibians can frequently be increased 10 to 15 mph through proper trim attitude. Of course, this is an extreme gain due to reducing flow separation around the hull chines and step, but such increases indicate the extent of drag reduction possible through proper attention to cruise trim.

Comparing drag

The methods used to determine drag values are of interest because drag is so basic to all aircraft performance, to say nothing of the importance of high cruising and top speeds to the sales department. The earliest experimenters determined drag values in a rather random manner with each pioneer using his or her own system of coefficients, making direct comparison difficult at best.

For a time, flat plate area was an established measurement for drag evaluation. This term referred to a flat shape of sufficient area to produce the same drag as the airplane would have in level flight; the smaller the area, the cleaner the airplane. Fortunately, during the early 1930s nondimensional performance coefficients became more common through development of increasingly sophisticated wind tunnel test techniques, and remain the standard for current work.

The sphere was considered a primary reference for determining comparative drag values because it was a precise shape and supposedly would provide the same drag reading, even when tunnel flow was not uniformly parallel to the test section walls. In fact, drag readings from a well-tested reference sphere became the basis for determining test section flow quality in newly constructed wind tunnels.

For the past 60 years, airplane drag values have been derived and compared on the basis of total wing area and maximum level flight speed. In accordance with aircraft engineering practice, total wing area includes the area between the spanwise projection of the wing leading and trailing edges as shown by Fig. 6-6.

Since we are primarily concerned with drag evaluation and reduction in this chapter, the study of 15 representative domestic and foreign private aircraft presented by Table 10-2 is of interest for comparison and design purposes. In addition to wing area and airplane total drag coefficient C_{D_T}, wing and power loadings as well as maximum speeds and gross weights have been included.

Table 10-2 Comparison data for domestic and foreign general aviation landplanes of similar power

Airplane model	Total drag coefficient C_{D_T}	V_{max} sea level, mph	Wing area S_w, sq ft	Gross wt W, lb	Horse-power hp	Wing loading W/S_w lb/sq ft	Power loading W/hp lb/hp	Total loading W_T
PIK-15 (Finland)	0.0383	142	150.7	1684	150	10.9	11.0	21.9
PA-18 Super Cub	0.0451	127	178.5	1750	150	10.0	11.6	21.6
M-B-B B0-209-150	0.0339	164	110.0	1808	150	16.4	12.1	28.5
Poulet DM-165	0.0305	165	132.4	1850	165	14.0	11.2	25.2
SAAB MFI-15B	0.0352	155	134.5	1930	160	14.4	12.1	26.5
ZLIN42	0.0556	136	141.5	2028	180	14.3	11.3	25.6
Maule M-4	0.0330	147	152.5	2100	145	13.8	14.5	28.3
PA-28-140 Cherokee	0.0384	139	160.0	2150	150	13.4	14.3	27.7
Socata Rallye 150	0.0566	130	132.4	2160	150	16.3	14.4	30.7
Grumman American Traveler	0.0350	150	140.0	2200	150	15.7	14.7	30.4
Beech B19 Sport	0.0412	140	146.0	2250	150	15.4	15.0	30.4
Cessna 172	0.0353	139	174.0	2300	150	13.2	15.3	28.5
Lark Commander	0.0419	138	180.0	2475	180	13.7	13.7	27.4
Aero Commander 112	0.0290	165	152.0	2550	180	16.8	14.2	31.0
Wassmer WA-41	0.0341	150	172.0	2645	180	15.4	14.7	30.1
Average	0.0389	146	150.0	2125	160	14.2	13.3	27.5

As noted in Fig. 6-8, the combined value of wing and power loadings W_T should be kept less than 28 for acceptable landplane performance. When designing seaplanes, the combined loading should be less than 26 for truly satisfactory water operation, but is usually nearer 30 due to greater structural weight of the hull or floats plus the useful load that must be carried to justify the power required. As a matter of interest, data for the popular Piper PA 28-140 Cherokee quite closely approximates the average values for all 15 aircraft of Table 10-2.

Airplane drag is compared at maximum level flight speed V_{max} because at that operational point powerplant thrust equals airplane total drag. As a result, an equilibrium speed is established that can be readily converted into the airplane nondimensional total drag coefficient based on wing area.

This relationship is shown by the point B intersection of the power-required and power-available curves of Fig. 10-1, with the corresponding airplane total drag coefficient value C_{D_T} determined by the following formula:

$$C_{D_T} = \frac{(\text{horsepower})(\text{prop efficiency})(146,620)}{(S_w)(V_{max_{mph}})^3}$$

where S_w is the wing area and prop efficiency is usually 0.75.

It is also possible to determine the airplane drag D, in pounds by using

$$D = \frac{(\text{horsepower})(\text{prop efficiency})(375)}{V_{max_{mph}}}$$

and then finding the value of C_{D_T} by the nondimensional formula

$$C_{D_T} = \frac{\text{Drag}_{lbs}}{(1.0756)(\rho)(S_w)(V_{max_{mph}})^2}$$

where ρ = sea level mass density of standard air = 0.002378.

Rather than become too deeply involved in performance calculations, it is sufficient to note that nondimensional coefficients let us find lift and drag values by using one basic type of formula.

The references at the end of this chapter contain a brief nomenclature list (1) and a glossary of formulas (2) used to determine aircraft performance, and speed conversion values (3), as well as a series of useful design references (4). Although this material also applies to other chapters, it is presented here for want of a better location. In addition, these references provide an excellent data source for those of you wishing to study performance calculation in greater detail.

Future performance

In summary, it can safely be noted that private aircraft performance has not experienced any "great leap forward" over the past 60 years. Although many technical and operational advances have been realized, greatly contributing to flight safety and reliability, our speeds are about where they were in the 1930s. During that period of grass airfields, the four-seat, streamlined fixed gear Percival Vega Gull of Fig. 10-2 cruised at 160 mph with a 200-hp Gipsy six engine, and offered a standard range of 600 mi. Produced in some quantity for that time, with 90 aircraft delivered, this design won the 1936 King's Cup Race against aircraft of similar performance.

10-2 *Percival Vega Gull.*

What has happened to our engineering over the past 60 years? We have slightly improved upon the Vega Gull's performance by retracting the landing gear on the 200-hp Mooney and Piper designs, but only Robin and Wassmer of France currently offer fixed gear aircraft approaching 1936 speeds.

There certainly is a design challenge to be met, and the Popular Flying Association of Great Britain is doing something to encourage useful development along these lines, while also adding a bit of sport to weekend rallys. This group has established a combined aircraft and pilot competition scoring system based on ROC, cruising miles per gallon, takeoff and landing distances, and the aircraft maximum/minimum speed ratio.

Aircraft are separated by categories, such as single-place, two-seat, four-place, and twin-engine types. Rate of climb is scored at 1 point per 100 ft/min; miles per gallon clocked over a 1-hr period are rated at 1 point per mile; takeoff distance is added to the landing run, with the sum divided by 2 and scored at −1 point per 100 ft; while the speed ratio of V_{max} divided by V_{min} is reduced by 1 and given 10 points per each remaining unit.

Such competition not only adds to the pure enjoyment of sport flying, but also improves pilot technique while encouraging some progressive thought and effort toward more advanced private aircraft development.

Let's try an example of this scoring to demonstrate superior aircraft performance. Let us first consider a four-place, single-engine design climbing 800 ft/min at gross weight, delivering 18 mi/gal, taking off in 800 ft and landing in 700 ft, with a maximum speed of 170 mph and stalling speed of 55 mph. This airplane would rate as follows:

$$\text{Climb: } \frac{800}{100} \times 1 = 8.0 \text{ points}$$

$$\text{Miles/gallon: } 18 \times 1 = 18.0$$

$$\text{Takeoff and landing: } \left(\frac{800 + 700}{2 \times 100}\right) \times (-1) = -7.5$$

$$\text{Speed ratio: } \left(\frac{170}{55} - 1\right) \times 10 = 20.9$$

$$\text{Design total} = 39.4 \text{ points}$$

Now suppose that through revised design, the climb rate could be increased to 1000 ft/min, the takeoff run reduced to 600 ft, landing distance to 500 ft, and stalling speed dropped to 50 mph, with fuel consumption and top speed unchanged.

$$\text{Climb} = 10.0 \text{ points}$$

$$\text{Miles/gallon: } 18 \times 1 = 18.0$$

$$\text{Takeoff and landing: } \left(\frac{600 + 500}{2 \times 100}\right) \times (-1) = -5.5$$

$$\text{Speed ratio: } \left(\frac{170}{50} - 1\right) \times 10 = 24.0$$

$$\text{Revised design total} = 46.5 \text{ points}$$

Note the weight given for the short-field performance parameters of stalling speed, takeoff run, climb, and landing distance. All of these translate into cross-country utility and flight safety. This is where our future operational challenge lies: increasing small-field performance without penalizing efficient cruising speed.

If all active pilot organizations sponsored sport-flying events of this type on a regional and national basis, I'm sure we'd see more interesting and rapid development of private aircraft design. Fortunately the Experimental Aircraft Association (EAA) is working toward that goal in the United States.

Superior powerplant installation, wing design, and weight control at current state-of-the-art levels would permit this improvement as shown by Table 10-1 summary results. As future private aircraft move toward such utility, we will eventually reachieve 1936 performance levels, but, unfortunately, at 1990s prices.

Today, the only personal aircraft reaching these performance levels are the home-built designs and kits developed by EAA members. Some of these aircraft are being approved for production by the FAA under the newly enacted Small Airplane Certification Compliance Program. Although the regulation only applies to two-place designs, new single-engine aircraft of 2700 pounds or less gross weight and carrying no more than four persons can be certified under recently enacted Part 21.24 as Primary Category Aircraft per chapter 17.

Endnotes

[1]Acronyms and Symbols. See chapter 15 for acronyms specifically related to seaplane design.

a.c.	Airfoil section aerodynamic center (approximately the center of lift)
AR	Aspect ratio = (span)^2/wing area = b^2/S_w
b	Wingspan in ft
BHP	Engine-rated brake horsepower = hp
c	Wing chord in ft or in.
C_{do}	Wing profile drag coefficient
C_D	Drag coefficient
C_{Di}	Induced drag = $(C_L)^2/\pi \cdot AR$
C_{D_T}	Airplane total drag coefficient
c.g.	Center of gravity
C_L	Wing section lift coefficient
C_{Mac}	Wing section moment coefficient
c.p.	Wing section center of pressure (center of lift)
D	Drag in lbs
g	Acceleration of gravity at 32.2 ft/sec^2
L	Wing lift in lbs (same as W for design work)
M_{ac}	Moment about the aerodynamic center
MAC	Mean aerodynamic chord (wing average chord)
psf	Pounds per square foot
q	Dynamic pressure in psf
ROC	Rate of climb in ft/min
RN	Reynolds number
S_{HT}	Horizontal tail area in sq ft
S_{VT}	Vertical tail area in sq ft
S_w	Wing area in sq ft
thp	Thrust horsepower (= engine power x prop. effic.)

V   Velocity (ft/sec, mph, knots)

V_{max} Maximum level flight speed (ft/sec, mph, knots)

V_s Stalling speed (ft/sec, mph, knots)

W   Weight in lbs (same as L for design work)

W_T The total of wing + power loading = $W/S_w + W/hp$

α   Angle of attack in degrees

β   Propeller blade angle

δ   Surface deflection in degrees

η   Propeller efficiency

ρ   Mass density of standard air:

$$\rho = 0.002378 \text{ @ sea level}$$
$$= 0.002177 \text{ @ } 3{,}000 \text{ ft}$$
$$= 0.002049 \text{ @ } 5{,}000 \text{ ft}$$
$$= 0.001928 \text{ @ } 7{,}000 \text{ ft}$$
$$= 0.001756 \text{ @ } 10{,}000 \text{ ft}$$

[2]Useful performance formulas:

$$C_L = \frac{W}{(\rho/2)(S_w)(V_{ft/sec})^2} = \frac{391\,W}{(S_w)(V_{mph})^2} \text{ at sea level}$$

$$C_{D_T} = \frac{(\text{horsepower})\,\eta\,(146620)}{(S_w)(V_{max_{mph}})^3} \text{ at sea level}$$

$$= \frac{\text{Drag}}{(1.0756)(\rho)(S_w)(V_{max_{mph}})^2} \text{ at any altitude}$$

$$D = C_{D_T}(\rho/2)(S_w)(V_{ft/sec})^2 \text{ lb}$$

$$L = C_L(\rho/2)(S_w)(V_{ft/sec})^2 \text{ lb}$$

$$M_{ac} = C_{Mac}(\rho/2)(S_w)(V_{ft/sec})^2 c \quad \text{ft-lb where c is wing chord in ft}$$

$$\text{MAC} = \frac{\text{wing area}}{\text{wing span}} \text{ ft (this is an acceptable approximation for basic design work)}$$

$$q = \frac{P(V_{fps})^2}{2} \text{ in psf}$$

$$\text{ROC} = \frac{(\text{thp}_{avail.})(33{,}000)}{W} \text{ ft/min}$$

$$RN = 6380 \times V_{fps} \times c_{ft} \times \frac{P_{alt}}{P_{SL}}$$

$$\text{thp} = \frac{D \times V_{ft/sec}}{550} = \frac{D \times V_{mph}}{375} = \frac{D \times \text{knots}}{326} \text{ or}$$

$$\text{thp} = (\text{engine-rated power})\,(\text{propeller efficiency})$$

$$V_2 = V_1 \sqrt[3]{\left(\frac{bp_2}{bp_1}\right)\left(\frac{\rho_1}{\rho_2}\right)} \text{ for variations in power with altitude}$$

$$V_{\max_{mph}} = \sqrt[3]{\frac{(BHP \times \eta)348.6}{(C_{D_T})\rho(S_w)}} \text{ where } \eta \text{ is usually near } 0.75$$

$$V_{S_{ft/sec}} = \sqrt{\frac{W}{(C_L)(\rho/2)(S_w)}}$$

$$V_{S_{mph}} = \sqrt{\frac{391 \times W}{C_L(S_w)}}$$

$$V_{S_2} = V_{S_1} \sqrt{\frac{W_2}{W_1}} \text{ for landing speed variation with weight of the same airplane}$$

[3]Speed Conversion Values
(ft/sec) (0.592) = knots
(mph) (0.868)   = knots
(ft/sec) (0.682) = mph
(knots) (1.1516) = mph
(mph) (1.467)   = ft/sec
(knots) (1.6897) = ft/sec

(a) D. B. Thurston, "Estimating Plane Performance via Comparison Method," *Aviation*, September 1946, pp. 55-62.
(b) W. Bailey Oswald, "General Formulas and Charts for the Calculation of Airplane Performance." *NACA Technical Report No. 408*, 1932.
(c) K. D. Wood, *Technical Aerodynamics*. McGraw-Hill Book Company, New York, 1935 and later editions.
(d) E. P. Warner, *Airplane Design Performance*. McGraw-Hill Book Company, New York, 1936 and later editions.
(e) Jacobs and Abbott, "Airfoil Section Data Obtained in the NACA Variable-Density Tunnel." *NACA Technical Report No. 669*, 1939.
(f) I. H. Abbott and A. E. von Doenhoff, *Theory of Wing Sections Including a Summary of Airfoil Data*. Dover Publications, Inc., New York, 1959.
(g) F. E. Weick, *Aircraft Propeller Design*. McGraw-Hill Book Company, New York.

11

Powerplant considerations

Although neither the airframes nor the engines of the earliest aircraft showed any degree of reliability, by 1916 to 1920 the structures of most production airplanes were superior to their engines. In fact, many of the Jennys and similar type aircraft used for barnstorming during the 1920s had such redundant structures that damage could literally be repaired with bailing wire and any type of wood available, or ignored completely until something fell off or collapsed from constant overloading.

We have come a long way since then. Modern engines and propellers are reliable and represent the major mechanical improvement in general aviation over the past 60 years. Although the problems resulting from discontinued 80/87 octane fuel tend to cloud the picture, engine development is rapidly adjusting to the higher lead content and aromatics used in available fuels. In fact, 1977 and later Cessna Skyhawks are equipped with a 160-hp Lycoming engine designed for use with present 100 octane fuels. While operational problems persist with older engines designed for lower octane ratings, engine manufacturers have developed flight and maintenance procedures for these engines to keep undesirable valve and plug deposits at a minimum.

In addition, the EAA and FAA have been cooperating to develop Supplemental Type Certificate (STC) approval for the use of automobile fuel in many production and home-built aircraft, depending on the type of flying intended.

Basically, the powerplant installation consists of the engine, propeller, and fuel and oil systems. However, to meet today's government requirements a fifth consideration must be added: reducing external noise levels. This seriously affects engine and propeller design.

Turbine engines are not covered in this book even though they enjoy a deserved reputation for reliability, plus up to 3000 hrs between overhaul (TBO). Technology does not permit production of even the smallest FAA certified turbines at a unit price comparable with the total cost of an equipped, four-place airplane, while fuel consumption tends to remain high; both these shortcomings preclude turbine use by the average private pilot. Until composite ceramics or similar materials can be used to cast or form turbine rotors and housings, it is unlikely that this type of engine can be priced to power standard production private aircraft.

Obviously, the light weight, reliability, and smooth operation offered by the turbine make it a most desirable power source once price and fuel consumption begin to approach reciprocating engine levels. And at the rate our current production reciprocating engines are increasing in price, this might not be too distant. Quite likely new materials, resulting price changes, and improved fuel consumption will eventually make turbine power more competitive with reciprocating engines. In the meantime, let's look at what there is to work with today.

Reciprocating engines

Required power is determined during the early stages of preliminary design, whether for a completely new airplane or a simple modification of an older model. Due to the extremely high investment necessary to obtain FAA approval for new engines, even when an existing design is being uprated or stretched, certified engines are not available in increments from 60- to 300-rated horsepower. Nor are the rotational speeds as low as necessary for reduced cabin and external noise levels. Nonetheless, the designer must work with available engines, while the engine manufacturers and aircraft companies must anticipate power ranges likely to be needed a few years hence. Fortunately, through such communication between the major small aircraft and engine manufacturers, suitable engines are usually developed and certified by the time they are needed for production aircraft.

Commercial transport manufacturers traditionally circumvent new engine problems by selecting an engine backed by military service. Departing from precedent, however, are the latest wide-body transports, all of which developed around new engines. The resulting service problems have been minimal, principally because proven military design features were used on each engine selected.

The worst situation for a designer is the combination of an experimental engine installed in an experimental airplane. Military aircraft development is constantly exposed to this costly and frustrating problem because both airframe and engine are joined at the limits of existing technology. Although this condition might be necessary to produce an advanced military vehicle, it must be avoided in private aircraft development to assure product reliability and reduce engineering costs.

Nothing can add to development and flight test expenses as fast as an unproven engine that doesn't cool or will not deliver rated power. Final flight work cannot be started until proper engine performance is realized, and only then can engineering begin tests necessary to develop acceptable handling characteristics and meet FAA certification requirements.

While it is most frustrating to face continuing development delays from experimental engine problems, an even worse situation is to experience repeated service difficulties from a newly certified engine installed in production airframes.

The excellent Continental Tiara engines suffered from an undeserved reputation arising from early manufacturing difficulties. This problem was amplified at the time because a few problem engines were installed in recently approved Piper Brave Ag planes being placed in service across the country.

Had these engines been proven in an existing and well-known design before being used on a new airplane, any engine problems would have been considered more routine. The aircraft designer should be constantly aware of this potential problem when selecting an engine for a new airplane, and never consider installing experimental engines in aircraft scheduled for production within the next couple of years.

With this overriding consideration, the few approved engines available in the desired power range must be carefully reviewed when finalizing any new airplane design. Such items as rpm, fuel consumption, engine weight, accessory provisions and access, type of engine mount required, and engine reputation with pilots and fixed-base operators further restrict engine selection within a very narrow field of candidates.

With today's pressures on fuel economy and operating efficiency, the designer must work to reduce airframe weight and drag in order to obtain maximum performance at minimum power. Cost is also paramount, since both engine price and airframe cost increase rapidly with power and weight, offering further incentives for efficient design.

While you might not have realized all this effort was necessary to produce that airplane you fly, these design options have been carefully and successfully analyzed for every major airplane type in production today.

The standard small aircraft engines presently available in the Western world are of 4- or 6-cylinder, horizontally opposed, reciprocating, air-cooled design. Known as *flat engines* because their principal dimensions lie in the horizontal plane rather than being vertical or circular (radial), various power and fuel system models are manufactured, principally by Continental and Lycoming.

Rolls-Royce produced a few Continental designs and modifications under license in England, and Franklin engine rights were purchased by Pezetel for manufacture in Poland. Pezetel has produced the 2-cylinder, 60-hp and the 4-cylinder, 125-hp models because these engines are useful for powered sailplane and efficient private aircraft designs.

As an approximation, flat engines with normal carburetion deliver about 0.50 hp/cu in of piston displacement. That is, a Continental 0-200 will be rated at 100 hp, a Lycoming 0-360 at 180 hp, etc.

Engine accessory provisions include fuel pumps, oil flow control and filter systems, self-starters, alternators with electrical output sufficient for the most complex IFR dual avionics installations, vacuum and hydraulic pump drives, governor drives for constant-speed propeller control, and on some Continental models, integral oil cooler installations. In addition, the fuel can be delivered to the cylinders through a gravity- or pump-fed carburetor, by pressure fuel injection, or by turbocharged air passing through an inline carburetor.

Of course, the fewer the accessory provisions, the cheaper the engine. Installation costs soar from the basic gravity feed to a turbocharged installation that can cost twice as much as the gravity feed system of the same sea level power rating.

Power vs. altitude

Practically all aircraft used for business, pleasure, and sport flying are powered by sea level-rated engines. That means the horsepower value assigned by the engine manufacturer is determined under standard sea level ambient air conditions of 59°F and 29.92 in of mercury. Because these engines are not supercharged they are also frequently referred to as normally aspirated engines, which is to say they breathe normally and without outside assistance.

Figure 11-1 illustrates that all normally aspirated engines, fuel-injected or not, decrease in power with altitude. However, note how the turbocharged TO-

11-1 *Power variation with altitude for some production engines using 100 octane fuel.*

360 holds its sea level power right up to 10,000 ft. With larger supercharge capacity, rated power can be carried to higher altitudes exceeding 15,000 ft. Because oxygen is required for sustained flight greater than 12,500 ft, and because most private flying is at altitudes less than 10,000 ft, the ability to carry full power to 10,000 ft is adequate for most flights and extremely helpful in maintaining a high ROC to cruise altitude. Further, the ability to develop full power in the less dense air of such altitudes means faster cruising speeds plus increased operational flexibility in marginal weather and IFR conditions. But at a penalty.

The penalty is increased empty weight compared with an engine installation of similar sea level power, greater initial and maintenance costs, and a higher level of flight technique. Note that the 210-hp turbocharged TO-360 weighs 22 lbs more than the fuel-injected 200-hp sea level-rated IO-360. It also costs 60% more. In addition, the TO-360 has a heavier and more complex exhaust system than the IO-360, as well as more sophisticated engine controls and control procedures requiring pilot attention.

As an alternative, the designer might consider installing a more powerful sea level-rated engine and restricting full power output until reaching some designated altitude. In effect, reserve cylinder capacity provides the engine with its own source of altitude performance. Figure 11-1 shows what would happen with an 0-540 engine rated at 250 hp at sea level. This installation could be limited to 200 hp at sea level by regulating manifold pressure through throttle control. But at 6500-ft pressure altitude, full throttle output is only 200 hp, and this drops off rapidly to 177 hp at 10,000 ft. In addition, the basic engine weighs 40 lbs more than the turbocharged TO-360, so the empty weight is greater, resulting in a climb penalty if the same useful load is carried. This is not a very effective trade-off, although the 250-hp 0-540 costs only two-thirds as much as the 210-hp turbocharged TO-360. So when altitude power is required, turbocharging is a more efficient design solution than excess sea level power.

Turbocharging provides some other important, although not so obvious operating advantages. Combustion air intake system losses normally waste 5 to 7% of sea level-rated power; by the time this air has entered the duct inlet, gone through the filter, and passed around a few duct elbows and bends, the sea level pressure has dropped 1.50 to 2 in. of mercury, even with some ram effect from flight speed. That means intake air pressure at the carburetor inlet will be around 27.50 to 28.00 in. of mercury instead of the sea level standard 29.92. Because turbocharging compresses the intake air, full sea level power can be maintained with turbo operation. This is an important gain providing superior sea level takeoff and climb performance, thanks to turbocharging.

In addition, the turbocharged engine results in quieter operation both inside and outside the cabin. This is a desirable design feature because it reduces pilot fatigue on long cross-country flights, while also making flying more acceptable to today's noise conscious public. Since turbocharged engines will be seen more frequently in the years ahead, a brief description of turbocharger operation follows.

Figure 11-2 is a greatly simplified diagram of turbocharged engine flow paths. First note that combustion air enters the intake duct just as it would for sea level carburetor or fuel injector equipped engines. The air then goes through an approved filter, past an alternate air inlet door that can be manually and/or automatically operated, and finally reaches the turbocharger compressor inlet. Incidentally, combustion air is usually referred to as *induction air*, with the delivery ducting and various related parts known as the induction system.

11-2 *Schematic of exhaust gas driven turbo supercharger with in-line carburetor. Fuel injection may also be used with metered fuel delivered directly into the intake manifolds.*

Once the induction air reaches the compressor stage of the turbo system, any similarity to sea level engine induction systems stops. In principle, engine exhaust gas drives a turbine that is coupled to a compressor. This compresser, in turn, takes the induction air and increases its pressure before metered fuel is added through a carburetor or injector system. The exhaust gas goes overboard after doing the work of compressing the combustion air. Of course, the fuel and air mixture burns in the cylinder, expanding to provide power through pressure exerted on the pistons, and in turn becomes exhaust gas working to compress the next air charge. This is why the turbosupercharger is so effective.

The turbosupercharger is sort of a perpetual motion machine. Engine exhaust compresses the inlet air to increase power, which in turn increases the exhaust flow, which in turn increases the amount of compression possible, etc. However, at some point the maximum structural design limits of the engine will be reached, preventing further increases in intake manifold pressure.

Excessive pressure buildup is controlled by limiting turbine power through waste gate control. As noted in Fig. 11-2, the waste gate is an overboard exhaust port bypassing the turbine. If the exhaust is allowed to flow overboard before it reaches the turbine, no power will be available to compress the intake air above ambient limits. As the waste gate is gradually closed with increasing altitude, more and more power becomes available to compress the less dense altitude air to desirable sea level limits (or even slightly greater manifold pressures if the engine is designed for such operation).

Waste gate position can be manually controlled by observing manifold pressure limits, or it can be automatically managed by a fairly complex system of barometric and oil pressure controls. The manual system is certainly the more simple, with some degree of pilot oversight taken care of by a built-in pressure-relief valve that operates at a few inches above desired limit pressure. And when it goes off, any pilot will remain alert for quite a period.

Manual control of turbocharger power is quite common on private aircraft, and a number of production powerplant installations are using this system. Since manifold pressures must be noted when operating higher-performance aircraft equipped with constant-speed propellers, little more effort is required to monitor and adjust turbine operation with altitude to maintain desired manifold pressure levels.

Older and larger engines such as the Pratt & Whitney R-985, R-1340, and up, had induction system impellers driven by the engine. These small-diameter compressors operated at very high speeds to increase combustion mixture pressures above sea level, and so provided a degree of built-in altitude performance. When high altitude power was desired, as for World War II fighter operation, the limits of the engine-driven impellers were extended by combining them with large turbochargers and intercooler radiators necessary to cool the huge amounts of compressed air. As a result, when the outer cowling was removed another solid wall appeared, but this time it was ducting, controls, supports, and baffles.

Fortunately, our small aircraft do not require such complex installations. The advantages of the separately mounted turbosupercharger in providing full power at sea level, power at altitude, and quieter exhaust due to the removal of energy during the compression work cycle, combine to make this accessory desirable for high-performance designs. As increasing numbers of production aircraft include turbocharging, the installed price will gradually lower to more acceptable limits; so expect to see turbopower become more commonplace in the future. In fact, to simplify powerplant installations, as well as reduce engine cost and weight, aircraft manufacturers should demand that turbocharger systems be built into higher-powered private aircraft engines.

Cooling requirements

Although the cowling should fit tightly to reduce air leakage and be of stream-lined and appealing design, it must have openings to supply the engine with cooling and combustion air, as shown in Fig. 11-3. The cylinder heads and bases, lubricating oil, alternator, and magnetos must receive adequate quantities of cooling air to function properly.

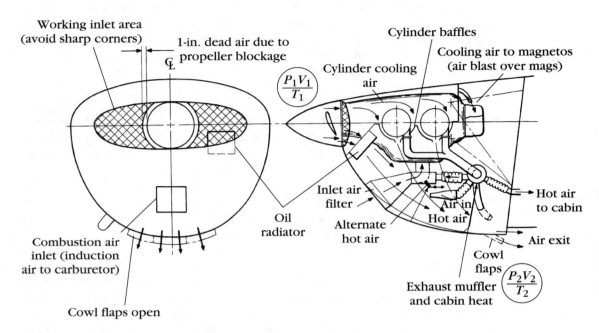

11-3 *Airflow through the powerplant installation.*

FAA test procedures require calibrated documentation of all these items during flight test, with corrections applied to assure cooling under 100°F hot day conditions. If the cowling openings are too large, air spills over and creates turbulent flow, adding to airplane drag. If the openings are too small, the cylinders and oil soon overheat. As a result, if the airflow volume ratio between climb and cruise flight is very large, that is, considerably more air is required for climb cooling than for cruise power conditions, cowl flaps should be used. Cooling is improved during climb by reducing cowling outlet pressure, which in turn drags more air over the cylinders; the cowl flaps are then closed to reduce cooling drag during cruising flight.

Reciprocating engines are cooled by a combination of air, oil, and fuel. As you know, air flowing over cylinders or through a radiator carries away heat. The oil radiator needs cooling air because the lubricating oil has picked up heat during its trip through the engine and must be cooled below 250°F to keep from breaking down into a useless, nonlubricating fluid.

The use of excess fuel is a classic, although extremely inefficient way to reduce cylinder temperatures. By enriching the carburetor fuel/air ratio jets beyond optimum, unburned fuel remains in the cylinder to cool the valves and heads after the combustion cycle has been completed. Unfortunately, this approach results in uneconomical fuel consumption plus high carbon monoxide levels in the exhaust. In order to properly define the correct fuel mixture ratio for each new airplane design, the engine manufacturer must specify the carburetor setting used for FAA certification flight tests. For all subsequent production aircraft, FAA specifications require the use of engines with an identical carburetor setting.

Engine cooling is also affected by manifold distribution, carburetor inlet airbox design, and many other small, but disturbing factors. While whole books have been written on the subject, I believe a simple cooling airflow example will serve our purposes. Engine installation specifications issued by the engine manufacturer list the volume of cooling airflow required over the cylinders, the permissible cylinder head and base temperatures, and the maximum oil temperature allowed. In order to protect their product liability exposure and engine warranty, the engine manufacturer as well as the FAA must approve the powerplant installation; so the cooling requirements must be well met and with some margin to spare.

Referring to Fig. 11-3, let us consider the installation of an engine requiring 2700 cu ft/min of airflow over the cylinders at full power. The crucial condition has been determined as a 70-mph climb speed at full power in 100°F hot day conditions. A 4 × 5 in. oil radiator with a 20 sq in rectangular flow area has been selected, with the radiator cooling air supplied from the cylinder air inlet opening. The following calculations provide flow areas of sufficient accuracy for the initial design approximation.

Calculating powerplant cooling air inlet area

Cooling air required = 2700 cu ft/min = 2700/60 = 45 cu ft/sec at standard conditions. At 70-mph climb, flow velocity = 70 × 88/60 = 102.7 ft/sec (where 88 = ft/sec for 60 mph, to convert mph to ft/sec). Assuming a 70% efficient cowling inlet design that results in a 0.7 cowling entrance airflow coefficient, the cylinder cooling air inlet area will be

$$\frac{45 \text{ cu. ft/sec}}{102.7 \text{ ft/sec} \times 0.7 \text{ entrance coef.}} = 0.626 \text{ sq ft}$$

$$\text{Oil cooler air inlet air} = \frac{20 \text{ sq in}}{0.7 \text{ coef.} \times 144} = 0.198 \text{ sq ft}$$

Total inlet area at standard conditions = 0.824 sq ft. To correct for a 100°F day, the air density ratio must be considered:

Standard density ρ = 0.002378
100°F density ρ' = 0.00221 (from density charts)

Inlet area for 100°F operation = $\dfrac{0.824 \times 0.002378}{0.00221}$ = 0.887 sq ft

0.887 sq ft = 128 sq in. of total cooling air inlet area

This area does not include combustion air volume or magneto and alternator cooling air. Note that the two density values given for sea level standard and 100°F days are the only ones required for most private aircraft powerplant design calculations.

To allow for inlet flow losses occurring around a propeller hub or spinner, the opening on each side of the cowling should be increased by starting the inlet area measurements 1 in. outboard of the spinner or hub diameter. Further, less-dense, high-altitude air cannot carry away as much heat as that at sea level; so for high-altitude work, additional cooling airflow might be required.

Calculating powerplant cooling air exit area

The outlet area can be based on the inlet area plus the expansion of cooling air passing over the hot cylinders and oil radiator. This calculation follows the simple gas law

$$\frac{P_1 V_1}{T_1} = \frac{P_2 V_2}{T_2}$$

where P_1 = inlet pressure in lb/sq in
P_2 = outlet pressure in lb/sq in
V_1 = inlet volume of air (or inlet area, for the same inlet and outlet flow velocities)
V_2 = outlet volume of air (or outlet area, for the same outlet and inlet flow velocities)
T_1 = inlet temperature in °F absolute (degrees Rankin) = ambient temperature in °F + 460°F
T_2 = outlet temperature in °F absolute = outlet temperature in °F + 460°F as shown in Fig. 11-3

Using for our example an inlet pressure of 14.6 psi, an outlet pressure of 14.1 psi, and a cooling air temperature increase of 100°F, for a 100°F hot day, the exit area becomes

$$V_2 = \frac{T_2 P_2 V_1}{T_1 P_2} = V_1 \left(\frac{T_2 P_1}{T_1 P_2} \right) = V_1 \left(\frac{660 \times 14.6}{560 \times 14.1} \right)$$

P_1 = 14.6;
P_2 = 14.1;
T_1 = 100 + 460 = 560°F;
T_2 = 100 + 100 + 460 = 660°F
    = 1.22V_1 (based on the same inlet and outlet flow velocities as noted above)

With the basic entrance area V_1 calculated as 128 sq in in climb, the corresponding outlet area V_2 = 1.22 × 128 = 156 sq in during climb. Another set of calculations is required for cruising flight, with the difference in outlet areas ideally provided by manually controlled cowl flaps to reduce cruising drag.

Although this is not an exact series of calculations, it is close enough for preliminary design work and for modifying your powerplant installation to accept a larger engine or a change in operating conditions. Remembering, of course, that any such modifications would be considered of major nature and must be substantiated by cylinder head temperature surveys taken in flight.

Air for combustion and accessory cooling can be picked up and directed from a NASA submerged air intake as shown in Fig. 11-4, which has been taken from NASA data. Based on drag studies and tests conducted by NACA and NASA in the mid-1940s and 1950s, this inlet has extremely low drag and lends itself readily to cowling design. Once the required air inlet area has been determined at the recommended width-to-height ratio of 3 to 5, the duct shape follows the simple formula and design in Fig. 11-4.

Alternate hot air

Some controlled hot as well as cold air is essential to safe powerplant operation. In order to prevent carburetor icing or to get rid of icing when it occurs, a source of hot induction air is required for both carburetor and fuel-injected engines. As shown on Fig. 11-3, alternate hot air is drawn from the exhaust heater muff directly into the carburetor when the hot-air control is moved to the "on" position. This control is usually grouped on the instrument panel with the other powerplant controls; its operation moves the carburetor airbox hot-air door to shut off the normal cold-air supply and admit hot air instead.

When alternate hot air is applied, the carburetor inlet air temperature rises sharply. In fact, FAA regulations require flight demonstration of a 90°F carburetor air temperature rise for sea level engines. This test must be performed in 30°F outside air free of visible moisture and with the engine operating at 75% power.

Because hot inlet air has less density than cold air, making the engine run excessively rich, applying heat causes some loss of engine power. The combination of this power loss plus flow losses through the alternate hot-air induction system cause the rpm to drop off, as will be noted during engine runup heat checks.

In order to keep the heater muff and hot-air delivery hoses from overheating, a small outlet hole is located in the airbox near the hot-air door to keep a small volume of heated air flowing through this system. Whenever possible, inspect the airbox as part of the preflight check to ensure the opening remains unblocked.

The turbocharged engine does not need an alternate hot-air source because intake air is well heated during its passage through the compressor section of the turbine. Even when the turbine is just idling over, the housing will provide sufficient heat rise to discourage icing. However, an alternate air source is still required because the inlet air filter could ice over. In this case, air can be automatically or manually selected from inside the cowling as shown in Fig. 11-2.

To use data:
- Select desired inlet width W
- Use aspect ratio of 3-5 for inlet opening (width ÷ height)
- Draw 7° angle from entrance H to skin line to obtain L
- Apply coordinate values to numbers obtained for L and W

Table of coordinates

X/L	$Y/\frac{W}{2}$
0	0
0.1	0.004
0.2	0.084
0.3	0.234
0.4	0.386
0.5	0.534
0.6	0.612
0.7	0.688
0.8	0.764
0.9	0.842
1.0	0.917

11-4 *NASA submerged air inlet permits a minimum drag, smooth cowling design.*

Simplifying operation

The wet sump oil system with engine oil located right below the cylinders has greatly simplified powerplant installations in small aircraft. The separate oil tank with connecting hoses was cumbersome, heavy, and expensive, although it is still required for aerobatic aircraft if the engine is to keep running during sustained inverted flight.

A few other features could be incorporated into the powerplant systems to improve flight safety while reducing opportunities for "pilot error."

A truly single-tank fuel system with a 5-gal or similar reserve tank, would be one such improvement. Although it is most difficult to get two or more tanks sufficiently interconnected to permit equal and uniform flow from each one, I'm sure that properly sized outlet and vent lines plus a feed sump could be developed to permit such simplified fuel control. Of course, a puncture of one tank or rupture of one of the outlet lines would mean the loss of all fuel. This situation could be readily overcome in a dual-tank system using a three-way valve to select tanks "one" or "two" independently, as well as "both" tanks.

The difference between a dual-tank system and most existing ones is that selecting "both" tanks would be standard, with tanks "one" or "two" turned on only when separate fuel quantity gauges or visual inspection indicated something was wrong.

In these days of increased fuel economy, a simple, cheap, and reliable automatic mixture control is most useful and ensures maximum duration for all cross-country flights. An improvement could be barometric pressure and/or exhaust gas temperature controlled, with manual override retained for pilot operation if necessary.

Another safety device deservedly receiving increased attention is a spring-loaded suck-in door located downstream from the induction air filter. This door opens automatically to provide induction air if the inlet air filter becomes iced over or otherwise blocked. This feature won't prevent or remove carburetor icing, so a source of alternate hot air is still necessary and must be provided for both carburetor and fuel-injected engines.

Some standardization of the cabin's location, size, knob contours, and knob colors for powerplant controls would be helpful. I prefer the quadrant arrangement over push-pull controls because they seem much easier to identify and more natural in operation, such as in the Cherokee installation. However, hiding the tachometer or manifold pressure gauge behind a control wheel does not contribute to reducing pilot error.

Powerplant instruments should be standard in size, calibration readout, and panel location relative to the powerplant and flight controls. As pilots begin to fly more than one type of (rental) airplane, the instrument and powerplant control differences become crucial. The FAA has somehow overlooked this and, it is hoped, the General Aviation Manufacturers Association (GAMA) will initiate suitable standards before the government feels obligated to do so.

The following color identification is recommended as a start toward powerplant controls standardization: throttle, black; mixture, red; and propeller pitch, blue. Hopefully, the industry will follow up by standardizing the other items mentioned.

Finally, oil radiators mounted directly on the engine crankcase eliminate the danger of broken oil lines and possible engine compartment fire. Although optional radiators of different cooling capacities would be required to cover different aircraft operating speeds, the simplified installation as well as the safety advantages make this small feature quite worthwhile, as Continental Motors has recognized by including integral oil coolers on some of their engine models.

Propeller design

There is nothing mysterious about propeller design or selection, with the possible exception of deciphering the numbering system all propeller manufacturers seem to assign to their products.

The first propellers were blades that were paddlelike in shape, made of wood, and of fixed-pitch design. That is, the blade angle could not be changed in flight. One notable exception was a variable pitch, cambered blade, aluminum alloy propeller developed in Italy by Crocco and Ricaldoni in 1907 and used to power a hydrofoil-supported boat at speeds exceeding 50 mph. This seems to have been an unexplained and temporary advance in propeller development, a real advancement on the progress curve, since little serious production effort was directed toward variable pitch propellers until propeller design theory and aircraft speeds made this improvement both possible and mandatory during the 1920s.

Today, most lower-performance aircraft have fixed-pitch propellers, although they are now made of forged aluminum alloy instead of wood while the higher-performance retractables have constant-speed props. Why this difference?

Figure 11-5 illustrates the relative thrust of four propellers plotted against airplane speed. Although these curves are approximately the same shape for comparison purposes, it is readily apparent that a constant-speed propeller encompasses the maximum thrust capabilities of the three fixed-pitch designs. This is because it can vary blade angle with airspeed to maintain selected rpm.

The fixed-pitch climb prop, used to obtain maximum climb thrust for towing sailplanes or for small seaplane operations, has a low blade angle. This is necessary because best climb is obtained at the low-speed end of the flight envelope and each blade advances forward through the air only a comparatively small distance per revolution; so the inclined plane of propeller advance through the air must be fixed at a small (blade) angle. But even the climb prop cannot be set at as low a blade angle as the constant-speed prop, because such a low fixed-pitch blade angle would overspeed the engine when the airplane increases from initial takeoff to climb speed. Because the constant-speed propeller blade angle increases with airspeed, the constant-speed propeller can be set to a lower blade angle than any fixed-pitch type to obtain greater static thrust for improved takeoff performance.

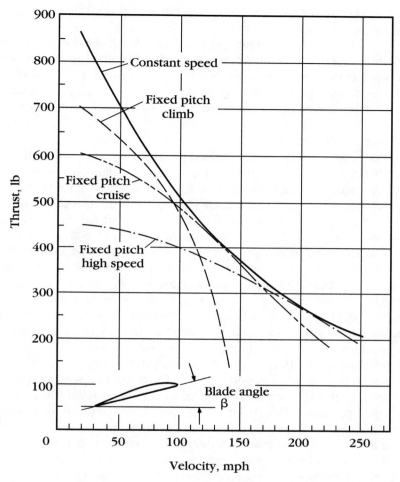

11-5 *Propeller thrust vs. velocity, constant speed and fixed pitch types.*

Further, when the fixed-pitch climb propeller equipped airplane levels off to cruise, the engine rpm reaches the red line long before maximum cruising speed is attained. This is caused by the effect of forward airflow velocity in unloading the blade and is the result of the low blade angle required for climb performance. Hence, you won't find many climb propellers on private aircraft.

A cruising propeller is the most common fixed-pitch type. This blade setting is a compromise between optimum climb performance and operation at or just below rpm red line at level flight maximum speed. As a result, static thrust is reduced in comparison with the climb propeller, as shown in Fig. 11-5, although at higher speeds the cruise propeller delivers more thrust because of its greater blade angle. The constant-speed propeller has superior takeoff and climb performance compared with the fixed-pitch cruising prop, but offers little advantage in the cruising speed regime.

In view of this, why not a two-position propeller instead of a constant-speed design? Many two-position models were offered following World War II, but their proper control required a governor; at which point it became obvious that a constant-speed type (which must have a governor) could be offered just as cheaply while delivering optimum performance throughout the entire airplane speed range. And so the successful two-position designs became production constant-speed propellers.

As a final fixed-pitch/constant-speed propeller comparison, high-speed prop characteristics have been included in Fig. 11-5. The relatively low static and slow speed thrust values result from high blade angle slippage at low forward speeds; the blade angle is so great that it just churns air around instead of efficiently displacing a mass of accelerated air. The early Cleveland Air Racers ran along the ground for such great distances because of the resulting low acceleration obtained from their high-speed props. In fact, new designs were frequently just able to take off with the propeller selected for racing performance airspeeds. Once airborne, high blade angle propellers increase in efficiency with airspeed.

Because of their ability to change blade angle with airspeed, constant-speed propellers can include high-performance operation as well as maximum takeoff, climb, and cruise performance. It is quite obvious why the constant-speed propeller is selected for modern higher-performance aircraft.

In another application, although certainly not a high-speed one, seaplane performance is greatly improved by use of a constant-speed propeller. Figure 11-6 illustrates the characteristic increase in static and takeoff thrust compared with a climb propeller. This thrust margin is extremely crucial for heavily loaded seaplanes and can make the difference between getting off the water or remaining a low-speed powerboat. Although the constant-speed propeller weighs and costs more, and will require some additional maintenance compared with a fixed-pitch type, the increased thrust and resulting reduction in takeoff run can be a vital safety feature in rough water operation.

This section has shown that constant-speed propellers should be used on all aircraft requiring maximum static thrust and cruising performance, regardless of the actual speed range. Whether the blades are fixed at a selected angle, or change pitch to match airplane velocity for constant-speed operation, the efficiency of the propeller blade section and planform will have a marked effect on airplane performance.

Figure 11-7 illustrates a typical production constant-speed propeller. Notice maximum efficiency occurs at cruising and higher speeds. Unfortunately, takeoff and climb occur at the low-speed end where efficiency is very poor. Because blade efficiency is an indication of the amount of engine brake horsepower converted into propulsive thrust, it is important to have propeller efficiency as high as possible throughout the entire speed range. For example, if the propeller efficiency can be raised 5% at an 80-mph climb velocity, the typical 2150-lb air-

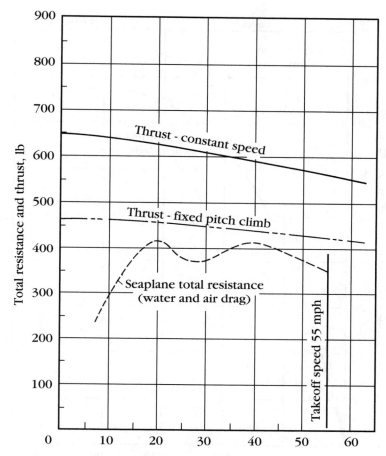

11-6 *Seaplane takeoff drag vs. propeller thrust.*

plane ROC will increase from 620 to 675 ft/min; this is about a 10% gain in climb for a 5% improvement in propeller efficiency.

Propeller operating efficiency is not entirely the responsibility of the propeller manufacturer. Notice that Fig. 11-7 shows two efficiency curves. The upper one is cheerfully submitted by the propeller designer; the lower curve shows what we usually end up with when the production installation has been completed—some 10% below possible maximum efficiency.

Where did the lost thrust disappear? It was taken a bit at a time by the thick blade root section necessary for structural stiffness and blade attachment, by the hub itself, by the cowling positioned close behind the inner blade section (or by the body turbulence ahead of a pusher installation), by tip flow losses, by cowling inlet flow losses, and by surface irregularities arising from manufacturing tolerances and service erosion.

11-7 *Propeller efficiency vs. velocity.*

Correcting or improving any of these factors will correspondingly increase propeller efficiency and airplane performance. The spinner is a step in that direction, although some payment is necessary in the form of increased weight and cost. So the airplane designer can materially improve operating efficiency through cowling refinement, short shaft extensions to locate air inlets further behind the propeller disc, Cessna style, and remote drive shafts that permit fairing the propeller spinner into a small-diameter, shaft-support structure.

Tackling the design-improvement problem from another direction, development work by all propeller manufacturers is currently underway to reduce tip losses. This research is intended not only to improve blade efficiency but also to lower propeller noise. Modern engine installations are becoming quieter through use of mufflers and turbochargers to such a degree that propeller noise is becoming increasingly predominant, as shown in Fig. 11-8. So EPA (Environmental Protection Agency) noise requirements are indirectly providing additional benefit in the form of decreased tip losses and increased thrust.

Blade efficiency is also being increased all along the span through application of supercritical airfoil sections to blade design. The Advanced Technology Light Twin (ATLIT) development sponsored by NASA was equipped with Robertson Aircraft/Pacific Propeller designed supercritical propellers as an initial step in developing a new blade concept for general aviation use. A marked improvement in ROC was realized, as shown in our previous example of increased climb resulting from increased propeller efficiency.

Propeller efficiency is also affected by the propeller activity factor, an empirical comparative design value that increases with blade width and diameter reduction. Propeller efficiency tends to decrease with the increase in activity factor necessary to deliver more power from a given diameter.

The subject of two- vs. three-bladed propellers frequently arises with an increase in power. In such cases, diameter is an important consideration. For ex-

11-8 *Propeller tip speed basic noise curve.* Hamilton Standard Data ample, study of a recent 250-hp constant-speed installation indicated that a three-blade 76-in diameter propeller was slightly more efficient than a 76-in diameter two-blade, and much more efficient than a 74-in diameter two-blade one, particularly at climb speed. However, an 80-in diameter two-bladed design was the most efficient of all throughout the entire flight range, provided such a large diameter could be used.

Increased blade area is required to absorb more power in much the same manner that increased wing area is desirable to support more aircraft weight; the greater the horsepower, the more blade area required for a given propeller diameter. Unfortunately, once a diameter restriction is imposed, an increase in blade area necessary to absorb increased power can result in too high an activity factor value with an accompanying loss in thrust horsepower. At this point the three-bladed design with a lower activity factor becomes desirable because it will be more efficient. And that means it will deliver more thrust horsepower for improved climb and cruising performance.

As a general rule, unless the larger diameter that is required is limited by tip speed or clearance problems, a two-bladed propeller will be lighter, cost less to purchase and maintain, and be as efficient as a three-bladed propeller capable of delivering the same engine power.

The continuing argument to resolve tractor vs. pusher propeller design advantages is basically concerned with the relative efficiency of these two configurations. Although many sound features can be advanced for either type, the practical maintenance aspects of aircraft operation must favor the tractor design. Any aft-mounted propeller is subject to damage from turbulent flow, materials coming from the cabin, loose hardware left inside the cowling, and debris thrown up by the landing gear. Further, propellers mounted at or above the wing trailing edge tend to reduce wing lift at the higher angles of attack associated with short-field takeoff, as well as during abrupt power corrections in the approach or landing configuration.

The fact that the Cessna 337 aft (pusher) propeller is capable of delivering greater single-engine climb than the forward one is not surprising because the rear constant-speed propeller is operating in a regime of reduced velocity airflow. Air is slowed by passing over the fuselage and, as shown by the comparison curves in Fig. 11-5, thrust increases with decreasing velocity. This condition was particularly apparent on the huge airship Graf Zeppelin where the slower speed boundary layer became so deep that the aft engine delivered 15% more measured thrust than expected from design calculations.

No propeller review would be complete without a few words concerning noise reduction. Not only is external noise of prime importance from the environmental standpoint, but also cabin noise remains one of the major reasons private flying does not enjoy broad public acceptance. It is not only a social barrier restricting normal conversation, but also extremely fatiguing as well.

Propeller noise is basically a function of tip speed; the nearer the tip speed comes to the speed of sound, the greater the noise level, as shown in Fig. 11-8.

Although a simple solution might seem to be to reduce rpms, this approach must be accomplished either by gearing the engine or using a powerplant of greater displacement turning at lower speed. Either way, the installation cost and weight both increase for identical power compared with a direct-drive, high-speed engine.

In addition, to convert the lower rpm horsepower to thrust requires greater blade area or diameter, or both, with the increased diameter again pushing up the tip speed. Although more blades could be added to absorb the greater torque for a given diameter, a two-bladed propeller becoming three-bladed for example, this solution also imposes an obvious cost and weight penalty as previously noted. In fact, it is not possible to reduce the noise level of any piece of efficient machinery without paying a penalty in cost, weight, or efficiency—or all three. Nonetheless, noise reduction is a necessary goal for increased acceptance of private aircraft by people both inside the cabin and on the ground.

With this objective, an effort to reduce propeller noise was mounted by Hamilton Standard with the Q-Fan concept, while Dowty Rotol conducted similar research in England with the Ducted Propulsor. Both of these developments use pressure ratio fans operating inside a ducted cowling. While there were some variations in cowling requirements, the major differences between the two designs were fan diameter, speed, and pressure ratio. As shown in Fig. 11-9, the Q-Fan was a smaller diameter, higher rpm, and greater pressure ratio unit than the Ducted Propulsor, with test data indicating the Ducted Propulsor was quieter than the Q-Fan at present stage of development. Both of these designs are heavier and more expensive than plain propellers of any type, and considerably more expensive than the standard fixed pitch, forged dural propeller that powers most small aircraft.

In the final analysis, for simplicity, weight, and price, the fixed-pitch propeller remains the best average compromise for most low-performance airplanes; and so will be with us as long as reciprocating engines power private aircraft. Readers interested in designing fixed-pitch propellers will find detailed procedures, calculations, and charts in chapter 10 endnotes.

Fuel system installation

A few brief notes about fuel systems should be adequate for our purposes, since actual fuel control is usually well covered in the owner's or airplane flight manual.

Safety should be the primary consideration for fuel system design. And this means no fuel in the fuselage or hull. Fuel should be stored in the wings with the inboard end of each tank at least a foot from the side of the fuselage to keep raw fuel clear of the cabin during any severe accident, and for lightning protection.

Of course, fuel located in the wings also reduces wing structural loading in flight in the same way wing weight does; both wing structure and fuel have downward acting weights that reduce the flight lift load (acting upward). For this reason, when two or more tanks are located in each wing panel, the inner

11-9 *(a) Ducted propeller configuration. Hamilton Standard "Q-Fan" installation, left-side view of pylon-mounted piston engine, Q-Fan propulsion pod. (b1) Ducted propeller configuration. Dowty Rotol "Ducted Propulsor" installation. Installation of ducted propulsor. (b2) Diagram of thrust recovery from cooling flow.*

tank of each side should be emptied first and the outboard tank emptied last. Using this procedure, the remaining outboard fuel weight will exert the greatest downward moment to reduce wing bending, as discussed in chapter 12. Note that carrying this to excess by locating all fuel in wing-tip tanks can be dangerous because in a spin condition weight concentrated so far outboard tends to wind an airplane into a fast, flat spin.

Fuel tanks should not be located in seaplane floats or down near the step region of amphibian or flying boat hulls. Such float and hull tanks could spray fuel around from striking a submerged object or during a hard landing, and sooner or later every amphibian should be expected to make a wheels-up land landing, which could grind the hull down to the step area "built-in-bomb" fuel tank. Again, the wings are an excellent and safer choice for fuel tank location.

One problem with a wing fuel tank installation is the relatively flat bottom surface of the tank resulting from wing dihedral rise. This tends to create a rather large residual fuel volume during FAA required slip and skid tests, as discussed in chapter 9.

It is possible to construct a baffle box at the fuel outlet/tank sump area that will hold about one gallon of fuel. This enclosure must be sealed to the tank lower surface and fed via a flapper valve that will admit fuel into the baffle box to feed the fuel outlet line, but will keep fuel from leaving the tank outlet box area during a slip or skid.

Since one gallon of fuel provides about 6–8 minutes of engine operation for the average personal airplane powerplant, that is more than sufficient time to satisfy the FAA slip and skid duration requirement, and so demonstrate that the airplane can operate with less than one gallon of residual fuel in the system. This is both a safety and an empty weight reduction feature, since all residual fuel must be included in the airplane empty weight.

As a note of caution, mud wasps can become a real problem if they build their nests inside fuel tank vent lines. It is possible to start the engine, warm up, and taxi to the active runway without noting any engine or fuel delivery problems. However, if mud wasps have been active, after about 5 min. of flight the fuel tank can collapse or the fuel pump quit because of diaphragm failure. And even a thorough preflight might not indicate a clogged vent because these insects can be packed in the mud an inch or two up from the vent outlet. Screened vent fairings would prevent such activity as well as related power failure accidents.

In addition to the convenience of fueling low tanks, being able to look into a tank during preflight to check the actual fuel level is a real safety item, as well as a source of personal relief. Unfortunately, many high-wing designs have no provision short of a step ladder for adequate fuel inspection, and this condition deserves further study and design improvement.

Current FAA design regulations require consideration of lightning protection for all private aircraft. Of course, the best protection is the total avoidance of thunderstorms. But because that is not always possible, grounded-flush-type filler caps and screened tank vent outlets will help. When the system permits, tank vents should be located under the wing near midchord to reduce the likelihood of secondary lightning strikes.

Various types of fuel delivery systems differ in complexity and cost, with a gravity flow carburetor installation being the least expensive. The fuel pump and carburetor combination are next, with the fuel-injected system being even more expensive. Because turbochargers can be used with either carburetor or fuel-injected engines, the fuel-injected turbosupercharger is the most complex and expensive of all fuel delivery systems. This is so not only because of the turbocharger itself, but because of the high-pressure pump and filter required to provide the pressure and clean fuel necessary for fuel injection, as well as the related automatic controls and bypass lines associated with fuel injection systems.

Fortunately, most of our flying does not require the complexity and maintenance cost of either the fuel injector or the turbosupercharger; a designer has little justification for using a fuel-injected, 180-hp installation when an equally powerful, lighter, and cheaper 180-hp carburetor-equipped engine is available. And this also becomes a safety consideration because the more complex an installation, the greater the opportunity for something to fail.

Reducing noise

Those of us who have flown for many years, and I'm sure this also applies to everyone interested in aviation, tend to overlook powerplant noise as something naturally associated with aircraft operation. Unfortunately, sometimes passengers are turned off by the combination of engine and radio noise, even when the radio is as low as it can be and still remain audible. By contrast, I admire the captain of an ocean liner or air transport, smug on the quiet bridge with barely perceptible engine vibration to indicate all is well. Without a doubt we'd all welcome the same low level of powerplant noise in the cabin.

Noise of any sort is tiring if experienced over an extended period of time. This fact partially explains the popularity of air-conditioned automobiles; they can be driven with all windows closed to exclude road noise, permitting long journeys with minimum fatigue. And I'd gladly substitute the quiet of the ocean liner bridge for the sound level of a modern large car, if ways could be found to offer such flight luxury. Although our aircraft save much time by being able to fly in a straight line, as well as travel quite a bit faster, I doubt if anyone can fly for 10 hr in any kind of general aviation airplane and not be thoroughly tired by the end of the journey. Yet it is possible to drive in air-conditioned isolation for 10 hr and arrive quite fresh. This is mentioned to indicate the need to eliminate another potential source of pilot error—fatigue; actually, cabin fatigue, resulting from long cross-country flights under constant engine and radio noise.

Aircraft noise annoyance is not limited to just the passengers. Airport and water-flying restriction suits have been brought to court by landowners across the country. Regardless of our personal interests, such restraints are understandable at present noise levels, with the end result being reductions in (1) aircraft safety from pilot fatigue and (2) utility from prohibited operations.

Fortunately, or unfortunately, depending on one's viewpoint, we are getting some assistance from the federal government in achieving noise reduction. Although present federal requirements are extremely complex, external sound levels for small propeller-driven aircraft will have to meet Fig. 11-10 decibel (dBA) values. So quieter flight will be realized, like it or not. However, you can be sure of seeing some redesign and cost increases.

11-10 *FAA noise standards for propeller-driven small airplanes.*

Sources of powerplant noise

Powerplant noise originates from two principal generators, the engine and the propeller. The propeller emits increasingly high and penetrating frequencies as tip speeds approach the speed of sound as shown in Fig. 11-8. Blade width and propeller solidity also contribute to propeller noise depending on disc power loading, while tip shape and section contour seem to have some effect as well. Outside of these factors there is little that can be done to reduce normal propeller noise, with tip speed being the most crucial consideration.

The engine, however, provides many opportunities for effective noise reduction. As shown in Fig. 11-11, based on Hamilton Standard research conducted for NASA and corrected to dBA values, energy is delivered from various engine sources besides the crankshaft. It is obvious that something should be done to reduce exhaust noise. And when the exhaust sound level is brought low enough, accessory drive, gear reduction, and other crankcase noises will begin to dominate the scene. In fact, only moderate exhaust noise suppression will bring gear and valve noises into focus.

Noise is also produced by accessory items such as the alternator, magnetos, fuel and vacuum pumps, propeller governor, and hydraulic pump that can rival the basic case noise. The entire accessory section and driven accessories are poorly positioned for the average airplane because their normal location is di-

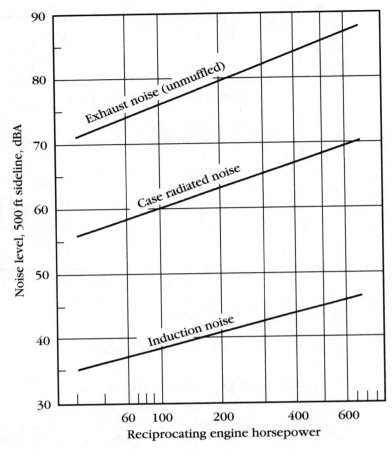

11-11 *Engine noise vs. power (approximate dBA values).*

rectly ahead of the firewall, permitting all accessory noise to be reflected from the engine right back into the cabin area. The engine manufacturers are finally recognizing this problem and locating more accessory items at the propeller end of the crankcase.

Induction noise is often of very high frequency due to small inlet openings and lead-in ducts, and so tends to influence the entire engine sound spectrum. Fortunately, very little energy is associated with induction flow into the normally aspirated (unsupercharged) engine, and so the flow frequency does not normally vibrate structural panels or other airframe components.

Suppressing powerplant noise

Certainly the engineering principle of eliminating a problem at the source applies doubly to powerplant noise reduction; the smaller the amount of energy emitted, the less suppression required to lower noise levels to acceptable values.

Looking first at propeller-tip speed noise, an obvious solution to decreasing energy from this source would be to reduce propeller diameter. However, as the diameter is reduced it might be necessary to add blades to absorb engine power, and so the weight and cost of the propeller both increase as previously noted. Another possibility is the reduction of propeller rpm. Of course, this approach also has power absorption problems, usually solved by increasing propeller diameter, and again climbing back up to higher tip speeds.

Hamilton Standard and Dowty Rotol research engineers were developing another solution to the tip speed problem as discussed under propeller design: a propeller of small diameter and high solidity driven at high rpm, but well below sonic tip speed. This was somewhat similar to a poor man's reciprocating powered turbine or bypass fan jet in principle. Although there is no doubt this design reduces internal and external sound levels, this gain is accompanied by increased weight and cost. As a result, these powerplant programs have been terminated for lack of interest, even though additional development might produce a configuration competitive with conventional propeller drive systems, but operating at a much lower noise level (see Fig. 11-9).

Current studies are underway to determine the relationship of tip shape and speed to emitted noise, while other propeller investigators are working on blade width and profile variations intended to reduce propeller noise at increased power output. Something must be done soon because propeller noise will become increasingly prominent as engines become quieter.

Returning to the engine and its principal sources of noise, mufflers are the obvious answer to the exhaust problem. Although they add weight and cost to the airplane, modern aircraft are being delivered with mufflers installed as original equipment. Unfortunately, automotive mufflers do not seem to work with aircraft engines, primarily because the automobile is normally operating at a small percentage of total power while the airplane is at or near maximum output. As a result, the combination of auto muffler and aero engine develops excessive back pressure, even with unacceptably large and heavy truck mufflers. It seems aircraft require their own specially-developed, low-back-pressure mufflers.

When turbocharging is used to obtain power at altitude, exhaust noise is lowered because of the energy removed by driving the turbine and induction air compressor. As noted a few pages back, turbocharging will find increased use on the next generation of private aircraft because it offers faster cruise at altitude plus quieter flight.

Accessory drive and engine case noise could be lowered by decreasing engine rpm. Of course, propeller tip noise could also be lowered, while engine and propeller size and weight would tend to increase for a given power output. Again, an engineering compromise and trade-off of some sort are necessary, including a completely new family of accessories. Improved gearing would also lower accessory and internal drive system noise.

Powerplant accessories

Before closing this chapter, I must mention three useful powerplant monitors: the cylinder head temperature, exhaust gas temperature, and carburetor icing indicators. All of these monitors contribute to operational safety as well as powerplant efficiency.

The cylinder head temperature indicator should be mounted on the hottest head as determined by flight test. Since the airplane manufacturer has demonstrated acceptable engine cooling under adverse conditions during the certification flight test program, one important reason to monitor cylinder head temperature, strangely enough, is to keep the engine from operating at too low a temperature. Low cylinder temperatures mean too much air is flowing through the engine, causing excessive cooling drag and probably inefficient fuel consumption as well.

Low cylinder head temperature might simply indicate the cowl flaps have been left open during cruise, or possibly that the cooling air inlet and/or outlet are too large in area. At any rate, excessive cooling is not good for the engine, and should be adjusted to more closely approach cylinder head limits as corrected to 100°F ambient conditions. In addition, cruising speed suffers when cooling drag is too high.

At the other end of the spectrum, a sudden increase in head temperature indicates that something has gone wrong and that power should be reduced at once. Possibly an exhaust stack developed a crack, a valve has gone, the cylinder wall has blown, or an injector has partially clogged. These are not comforting prospects, but early recognition of trouble provides time to find an acceptable landing spot.

The exhaust gas temperature indicator is of value when fine leaning is necessary to extend cruising range by operating at peak fuel consumption efficiency. Leaning in this fashion should be accompanied by head temperature monitoring to assure that fuel distribution does not change sufficiently to raise any one cylinder much above the average temperatures. By leaning to optimum recommended exhaust gas temperature it is possible to decrease fuel consumption to such a degree that the manufacturer's advertised fuel consumption rates can actually be realized, and this is a feat in itself.

Although carburetor icing can be a problem at any latitude, it is particularly disturbing in northern climates at almost any time of the year. ARP Industries of Huntington, New York has developed a patented indicator that among other duties signals by red dash light when icing conditions develop in the carburetor throat. This warning indicates the need for carburetor heat in sufficient time to prevent icing buildup. This is certainly an important safety device for powerplant induction systems prone to carburetor icing, which is why this manufacturer has been mentioned.

While many of us have flown for many years without benefit of these aids, their installation could make life a bit more pleasant, and safer as well. Considering the reliability of current production engines and propellers, the major requirements for an acceptable powerplant installation are:

- A well designed and tested fuel delivery system.
- A properly filtered air induction system incorporating an automatically operated emergency air intake door.
- An adequate source of alternate hot induction air to prevent or remove carburetor icing.

All other design considerations reviewed relate to operating efficiency, noise reduction, or pilot acceptance. As such they are important powerplant criteria, but must take a secondary position to the preceding three primary safety features.

12

Airframe design

A text of this scope cannot possibly cover airframe design in great detail. Because of this, significant criteria and load conditions are presented only for major components, with particular emphasis placed on the different methods and materials used for structural design.

Early aircraft were fabricated from bamboo, wood, steel tubing, and wire, not only because these materials were readily available, but also because high-strength aluminum and steel alloys had not been developed. Although the German Zepplin LZ-1 of 1900 used formed aluminum sheet for the girders and gondola cars, some time passed before the cost and strength of aluminum alloy allowed it to replace wood, mild steel, and fabric as an airframe structural material.

Today, practically all production aircraft are of sheet metal construction. This type of structure has evolved partially because the quality-control standards applied to aluminum alloy sheet, bar, extrusion, and tubing assure repeatability of material thickness and strength. In addition, the load transfer across sheet metals riveted and bolted joints can be calculated very precisely with existing analysis procedures, which helps keep ulcers under control. Sheet metal assembly also requires less time and less specialized skills than welded steel and fabric construction. Although wood is easily worked, the resulting structure tends to be heavy and subject to deterioration from prolonged exposure to weather.

Sheet metal aircraft manufacture requires a considerable investment in heavy machinery, tooling, and equipment as well as great quantities of generally skilled hand labor. And to magnify the cost problem, the limited volume of aircraft production defies application of known automation techniques. No automobile manufacturer would consider a single run limited to 15,000 units, yet that was the total United States general aviation production for 1976, divided among 10 different companies. Even though three manufacturers produced about 90% of this total volume, the largest individual quantities were still comparatively small when the various models are considered. Unfortunately, 1990s production is about one-tenth that of 1976.

Because aircraft cost is of interest to everyone and is directly related to airframe design, let us follow through the procedure establishing the market price of a new four-place, retractable-tricycle gear airplane. In the process, we will review the rather abstract production learning curve that is frequently mentioned but never defined when discussing aircraft pricing.

Cost vs. quantity

Very early in any cost study, the type of airframe structure used for a new design will be compared with those of prior production models in order to estimate manufacturing labor-hours requirements. For the average low-to-medium-performance sheet metal airplane produced in this country, the parts and assembly time for the one-hundredth unit will total approximately one labor-hour per pound of manufactured airframe weight. Because this weight does not include the engine, propeller, wheels, brakes, tires, and major hardware items, these components are deducted from the basic empty weight to establish the manufactured airframe weight for costing purposes.

For our example, engineering has determined the airplane empty weight to be 1350 lbs. The weight of the manufactured airframe then becomes:

Basic weight empty		1350#
Engine weight (equipped)	300#	
Propeller weight (constant speed)	50#	
Wheels, brakes, tires and tubes	40#	
Major hardware items	50#	−440#
Net weight manufactured		910#

From this point on, the airframe parts and assembly cost estimates are established by the learning curve. This is a logarithmic line plotting aircraft volume vs. manufacturing labor-hours to give the probable time required to build each airplane on a continuing production basis. For our relatively simple new design, production is estimated to move along an 85% learning curve as shown in Fig. 12-1, which has been prepared to cover the first 100 aircraft.

The 85% value refers to a production improvement rate in which each succeeding airplane requires only 85% of the time needed to complete the preceding one. If the airframe had been much more complex and estimated at many more labor-hours, production experience might have indicated a more rapid decrease in labor-hours as improved tooling became available. In this case, the learning curve might have been established on a steeper 80% slope basis, as given in the unit cost table of Fig. 12-1.

Our particular unit cost curve has been established by spotting 910 labor-hours at the one-hundredth airplane, and running an 85% learning curve through that point. The slopes for the normally used unit cost lines and the factors for the closely related cumulative average cost lines are given in Fig. 12-1 for use with 2 × 2 or 2 × 3 cycle logarithmic paper.

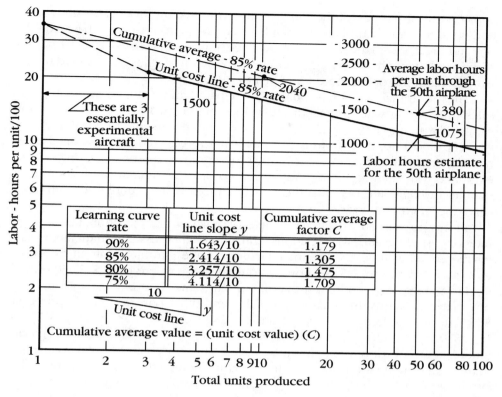

12-1 *A typical production learning curve and values for plotting different learning rates.*

Although the actual production hours estimated for any unit lie on the unit cost line, it is an unfortunate fact of life that all aircraft built up to that point have incurred labor-hours costs indicated by the cumulative average line. The cumulative average line value at any production point represents the average number of labor-hours per airplane required to produce aircraft up through that unit. For example, at the tenth airplane the 85% cumulative average line indicates that 2040 hours will be the average labor expended per plane for the first 10 units produced; that is, 20,400 labor-hours will be required to produce the first 10 aircraft.

Because learning curves are based on estimates, they should be considered an approximate indication of aircraft production time requirements. But they are the best guide there is and they do provide historical records for future manufacturing estimates if kept up-to-date as production develops.

The fiftieth airplane has 1075 labor-hours of unit cost line time scheduled for that airplane; yet the company supported production buildup costs averaging 1380 labor-hours per airplane up through the fiftieth unit, as shown by the cumulative average line. The difference is (305 × 50)= 15,250 labor-hours, and someone must pay for all the time expended in getting production under way and up to the fiftieth airplane. Remember that pricing estimates are based on the one-hundredth airplane produced, so even more labor costs have to be absorbed following the fiftieth unit before the program reaches an assumed break-even point at the one-hundredth airplane.

These extra labor-hours should be anticipated and included in the airplane sales price along with engineering development costs. If this is not done, the capital invested to develop model no. 1 will not be available to develop a subsequent model no. 2. Unfortunately, this is not recognized by most small engineering-oriented operations, and is one of the primary reasons they drop out of existence with such regularity. Any group considering aircraft production must plan accordingly and obtain sufficient capital to stay alive until sales deliveries can support current costs.

At this point in our cost review, the economics of plant location, taxes, wage rates, invested capital, labor relations, and a number of less important factors enter the equation. Because the manufacturing wage rate including all overhead can vary from $25 to $35 per hour, it could cost anywhere from $22,750 ($25 × 910) to $31,850 ($35 × 910) just to build this airframe. That is some spread. With raw materials, hardware, engine, propeller, and other purchased parts priced at $35,500, our airplane could cost from $58,250 ($22,750 + $35,500) to $67,350 ($31,850 + $35,500) to produce without including any manufacturing profit, amortization of development and production costs, sales costs, or dealer markup. Assuming a total wage rate of $30 including all overhead charges, the production cost of this airframe at the one-hundredth unit would be:

$$\$30 \times 910\# = \$27,300 \text{ assembled airframe cost}$$
$$\underline{35,500} \text{ purchased parts cost}$$
$$\text{Production cost} = \$62,800 \text{ at 100th unit}$$

Assuming the design, tooling, and production startup costs totaled $3,000,000 to be amortized over 500 aircraft, another $6000 per airplane must be added before any profit or dealer markup enters the picture. If such costs were amortized over only 200 airplanes, the base price would jump a prohibitive $15,000 over production cost; so such rapid development write-off is rarely attempted with small private aircraft. But how many second and third-level lightplane manufacturers ever produce even 200 units of a new model? And frequent nonrecovery of invested capital is one reason bankers do not care for the private aircraft industry, and also another reason why undercapitalized new airplane companies go out of business after a relatively short period of operation.

Getting back to the final price for our new four-place design, carrying the introductory costs over 500 units results in:

Production cost	$62,800
Development amortization	6,000
Subtotal	$68,800

In order to stay in business, the manufacturer must realize sufficient profit to satisfy the shareholders, provide for plant and equipment expansion and replacement, and have funds for new product development. Also, if this new design is to sell, it must be advertised. Experience has shown that advertising and sales expense runs about 10% per unit. On this basis our exfactory price becomes

Subtotal price	$68,800
Advertising costs	6,880
	$75,680
Profit at 15% pretax	11,350
Price exfactory	$87,030

Dealers need sufficient margin to permit some trading or list price reduction, so their share of the total is usually in the vicinity of 20 to 25% of list price, depending on the airplane price bracket and parent company policy. If the factory price for a dealer discount of 25% off list price is adjusted, the result is:

$$\text{List price} = \frac{\text{exfactory price}}{.75} = \frac{\$87,030}{.75}$$

List price offered through the dealer = $116,040

It didn't take long to get from the $62,800 production cost to the $116,040 list price, did it? And any avionics, IFR instruments, or special equipment options haven't been included. As you can see, sales list price is about twice production cost.

These are very reasonable production costs under existing economic conditions and will differ very slightly on a worldwide basis as a glance at current Beech, Mooney, Piper, and Socata prices confirm. Although it might seem expedient to subcontract the airframe to a comparatively low labor rate country, by the time necessary materials are imported into that country, the people there finally learn what they are supposed to do, and the airframe is shipped back into this country, the cost will be just as high as with domestic manufacture—but the quality will be much lower.

Low technology labor cannot reduce prices, but we can lower costs through a simplified design approach reducing the number of structural parts. The design challenge then lies in simplifying the airframe without unduly penalizing either performance or useful load. Let us see what can be done.

Cost vs. design

Seeing the word *cost* appear in a subheading for the second time might make you wonder if I am not straying far afield from airframe design. Such is not the case, since airplane cost is directly related to airframe complexity, as just shown in the production pricing example.

The lightest airframe for a given set of load conditions will have light skins plus many supporting ribs, stringers (skin stiffeners), and frames. It will also be the most costly structure to tool and produce, and unless well-tooled and carefully built the skins will appear full of wrinkles at rest and in flight. These slight surface variations can be eliminated by increasing skin thickness. The weight added by heavier skins can be somewhat decreased by increasing rib and stiffener spacing, and so reducing the number of airframe parts.

This design philosophy applies to all small aircraft produced in quantity in the United States, and also extends to modern military aircraft where the skin plating can be so thick it is machined to fit the airframe contour. Using wing structure as an example, you'll find that updated versions of old two-spar, fabric-covered wings might retain both spars even though redesigned for metal covering. The modern trend is toward a single beam wing having a torque tube leading edge and a light rear spar that picks up the wing torque and drag loads. This type of stressed-skin wing construction is shown in Fig. 12-2 and is ideal for integral leading edge fuel tanks, as also used on the similarly designed Cherokee series. It has a minimum number of detailed parts, and a simple three-bolt attachment or similar connection to the fuselage or hull.

A version of this type of wing structure is also used on Commander aircraft, which have been designed to FAR Part 23, amendment 7. These FAA design specifications require that at cruising speed a wing must be able to support 75%

12-2 *Stressed-skin wing construction used on Teal amphibian. (Based upon Flight International drawing.)*

of the crucial limit load after failure of a principal structural element and that it be able to do so with at least a 15% load safety margin.[1]

To meet the fail-safe requirement, it is essential that efficient alternate load paths be made available, as shown by the diagonal beam located at the leading edge root section in Fig. 12-3. If the main beam root fitting should fail, load would pass along the leading edge beam and be taken out by a leading edge-to-fuselage attachment; all of which adds many parts back into our simplified structure.

12-3 *The fail-safe wing structure of the Model 114B Commander, showing the alternate load path provided by the inboard leading edge beam.* Commander Aircraft

The wing drag load (acting forward or aft) is usually picked up at the fuselage from the inboard end of a small beam located near the wing trailing edge, as shown in Figs. 12-2 and 12-3. It does not require an additional leading edge attachment (which has been added just to satisfy the fail-safe design condition).

The fail-safe requirement should not apply to aircraft that are used only 200 to 500 hrs a year, and rarely, if ever, experience a 3 g flight load condition. Analyzing a similar structure for the Teal amphibian, indicates a fatigue life expectancy of more than 15,000 hr if the structure were loaded to 2 g every minute for the entire 15,000-hr life of the airplane. At an annual use of 300 hrs under such turbulent conditions, the airframe would be expected to last for 50 years, showing how ridiculous this FAA transport design requirement becomes when applied to general aviation use. The entire airframe would probably corrode long before fatigue could affect the principal structural members, while the owner would probably also wear out during the same 50 years.

Despite such unnecessary and costly requirements, a simple wing can be developed around very few parts by designing with stressed-skin construction. And the fewer the wing parts to be built and assembled, the fewer the labor-hours required, resulting in a less costly airplane as previously shown.

This design approach obviously applies to the fuselage and tail surfaces as well, explaining the relatively great distances between frames and ribs of most aircraft. But the point to remember is that the lightest structure requires many parts and much labor, while a practical structure of affordable price will have fewer parts and a somewhat greater airframe weight.

Fuselage design

The fuselage should be designed to satisfy two major criteria: (1) protect the passengers in the event of a crash, and (2) efficiently tie together the power-plant, wing, landing gear, and tail surface loads. Of course this must be accomplished with great interior volume for passenger comfort and minimum frontal area and contour drag for maximum performance, another set of challenging compromises for the design team to solve.

Probably the safest and most rugged passenger capsule is provided by the amphibian as a by-product of the hull bottom design load conditions. Since the special requirements of seaplane design is covered in chapter 15, it is not pursued here.

Ideally, passengers should be protected by a structure that will withstand a forward loading of 40 g based on the airframe weight. Metal sailplanes designed to this criteria have saved the lives of pilots who would have been killed by similar crashes in fiberglass fuselages. As an added consideration, if wings are designed to fail forward around the 20 g impact level, the fuselage structure need not absorb wing weight energy at impacts greater than 20 g, providing a form of inherent load relief that will reduce protective structure weight.

If 40 g protection cannot be realized without paying an extreme weight penalty, some reduced level of crash safety should be provided, extending down to 20 g as a lower limit. Shoulder harnesses that can definitely support a 100-lb upper torso at 40 g should also be provided, and should be anchored into some portion of the structure that can actually pick up and distribute 4000 lbs into the fuselage.[2] Chapter 13 discusses seat belt and harness restraint systems at greater length from the equipment installation standpoint.

Reduction of passenger fatalities resulting from off-field landings has received increased attention from both FAA and NASA. As a result, FAR Part 23, amendment 23-36 (change 24) was issued effective September 14, 1988 with paragraphs 23.561 and 23.562 revised to cover emergency landing impact and dynamics loading conditions relating to both airframe structure and occupants. These conditions are quite severe and generally require controlled failure of the seat and supporting structure in order to reduce deceleration loadings to acceptable values.

Fortunately, it is possible to simulate airframe structure and crash results by use of computer-based analytical models.[3] NASA, in association with the FAA, has crash tested airframes of modern design under controlled conditions, with details of each impact recorded through multiple instrumentation and camera locations. At contact angles as low as 15° with the ground and at 30-mph impact velocity, crucial parts of instrumented passenger dummies yielded whip loadings approaching 24 g vertically, but only about half that magnitude in the horizontal direction. Presumably the horizontal g values climb rapidly as impact angles and velocities are increased.

Detailed studies of personal and agricultural aircraft crashes based on the ratio of fatal accidents to total accidents have shown that the least safe personal airplane had 3.46 times the fatality record of the most crashworthy design, while the worst Ag plane had 14.5 times the fatality ratio of the safest model currently available.[4] With a spread of this magnitude, it is apparent there is some way to go before adequate crash safety is incorporated into fuselage design.

Surprisingly, little difference in crash survival could be determined between high- and low-wing aircraft—with the high-wing type actually slightly safer. But this finding could be the result of a trade-off, since most high-speed personal aircraft are of low-wing configuration and a greater percentage of high-speed aircraft accidents are fatal.

As a basic criterion, the safe fuselage design includes a passenger cabin capsule surrounded by energy-absorbing structure capable of supporting the passenger seat belt and shoulder harness restraint systems during impact.

Let's move on to fuselage design load conditions. As briefly noted in chapter 7, the larger the vertical and horizontal tail surfaces, the greater the aft fuselage loading and structural weight. Figure 12-4 is a greatly simplified presentation of this tail loading condition, but it helps to show how aft fuselage loads are generated. When the horizontal tail is loaded symmetrically as shown, that is, with the same amount of load on each side, the fuselage tends to bend down as loaded

12-4 *A simplified diagram of aft fuselage loading from horizontal tail load, aft fuselage weight, and vertical tail weight.*

in our diagram; and the further this load travels forward from the tail, the greater the bending moment because the arm distance to the tail load increases (moment = arm times load).

This bending must be resisted by the fuselage aft structure, and the further apart the supporting structural members or longerons can be placed, the lower the resisting load becomes. But as the fuselage is made deeper to reduce bending loads and so lighten the longerons, the skin area increases and becomes heavier. Some design study is usually advisable to review combinations of skin, stringer, and longeron sizes for weight optimization, but might not be particularly crucial because passenger accommodation will probably establish the cabin and aft fuselage shape regardless of load requirements.

In addition to vertical bending, our illustration also contains vertical shear. Shear of this type is probably best understood as a load acting as though it could be applied at every point along the aft fuselage. That is, the horizontal tail load acts with the same (shearing action) load value all along the aft fuselage right up to the wing beam. For our purposes, the shear load is supposedly balanced out at that point by countering shear forces acting downward ahead of the beam, as provided by the powerplant, forward structure, passengers, and equipment.

Of course, more than just the horizontal tail load is acting down, since the weight of the fuselage and tail surfaces also acts down for the condition shown and must be added to the horizontal tail load. The structural weights must be multiplied by the maneuvering load factor to obtain the actual design load, increasing both the bending moment and shear loads for this design condition.

All this might seem complicated at first glance, but is not if taken one step at a time. You will be pleased to know this load review does not become involved with detailed calculations better left to structural design manuals. Be aware, however, that other tail load conditions compound fuselage analysis.

Note the view in Fig. 12-4 looking forward. It contains a vertical tail loading that could result from a horizontal gust or maneuvering condition. Assuming the pilot applied up elevator and right rudder at the same time to effect a climbing right turn, the loading would be as shown. Just as the horizontal tail down load caused aft fuselage bending and vertical shear, the vertical tail load shown would result in the application of horizontal bending and shear in addition to the vertical loadings previously discussed.

Although the combined effect of vertical and horizontal tail loadings on the fuselage need not be investigated for so-called conventional tail locations having the horizontal tail positioned on the fuselage, this interaction must be analyzed for the T-tail, as noted in chapter 7. The resulting load conditions place the aft fuselage in torsion as well as bending, since the vertical tail center of pressure is located well above the fuselage. Fuselage torsional loading can be further increased by the unsymmetrical horizontal tail load condition required by FAA design specifications.

Figure 12-4 shows how fuselage torsion develops from a partial load on one side of the horizontal tail, full load on the other side, and a loaded vertical tail.

Of course, these tail loads can act up or down and left or right, so all possibilities must be investigated.

While I won't go into repetitive detail, similar bending occurs in the forward fuselage through application of vertical and horizontal design load factors to the powerplant, with torsion introduced into the engine mount and forward fuselage structure from engine torque. In addition, the nosewheel applies vertical and lateral bending and shear, plus torque, onto the structure; and it can develop all these loadings at the same time during a sideslip nosewheel contact. So land with the nosewheel clear for your own safety and to preserve the forward fuselage structure.

The wing carries bending and shear loads into the fuselage where they must be taken out through strong connections and balanced by inertia forces. This is why a load carry-through beam or set of compression tubes is carried across the fuselage, frequently just where the seats should be located. The wing attachment and fuselage carry-through structure must also be designed for an unsymmetrical wing loading of 30% maximum flight load: that is, full lift load applied on one side of the fuselage and 70% of that load on the other side.[5] This condition, which tends to load and rack the fuselage in torsion at the wing attachment points, must be resisted by the fuselage carry-through beam and then transferred into the fuselage structure.

Typical examples of modern fuselage skin and stringer construction are illustrated in Fig. 12-5. Figure 12-5a shows the reinforcement required around a door cutout in longeron-type and stringer-type structure, while Fig. 12-5b presents an advanced type of aft fuselage design. Note that this fuselage is assembled in halves joined at the top and bottom centerline. The frames are rolled integral with the skin, with light stiffeners providing skin panel support between frames. Although considerable tooling would be required to make such a simplified assembly, this design approach is all the more remarkable because it was developed and used on the Messerschmitt ME 109 and 110 fighters back in 1935–1938.

A similar structure was used on the two-place Ercoupe designed in this country by Fred Weick in the late 1930s and produced in quantity right after World War II. Nothing simpler has appeared on the scene since that time, and cannot, because the structural configuration of Fig. 12-5b contains only the essential parts required to carry the bending, torsional, and shear loads applied by the tail surfaces.

While on the subject of fuselage design, a note or two about windshields and cabin entry might be useful. In order to reduce aerodynamic noise, the windshield should be faired into its supporting structure, with the area around and aft of the windshield designed with smoothly curved contours. Air leaks should be well sealed because they are another source of noise as well as increased drag. Propeller noise can be considerably reduced by a thick windshield, which also acts as its own damper to reduce or eliminate noise originating from thin windshield vibration. Of course, cost and weight penalties accompany heavy glass, but they can be reduced somewhat by eliminating the

12-5 *Examples of well-designed fuselage structures. (a) Typical longeron and stringer construction showing reinforcement around a door opening. (b) An excellent example of simplified design. Note that the frames are formed integrally with the skin.*

center post that is not required to support a thick windshield, and so forward visibility is also improved.

A lightly tinted windshield glass will reduce glare and solar radiation, with such tinting particularly desirable for overhead viewing panels. The tinting must not be too dark, however, because night vision will be reduced to unacceptable levels.

Cabin access seems to be a constant problem with small aircraft. Whether high-wing or low, a tangle of feet, charts, and baggage seems to occur at each end of every flight. It seems strange that more attention is not directed toward solving this problem. Probably the Cardinal offered the best entry of any production private airplane design, but all other door-equipped models are quite clumsy.

One solution to the entry problem that is frequently used in Europe and by Grumman American in this country is the sliding canopy shown in Fig. 12-6. This canopy can be moved back to ventilate the cabin on the ground, offers the safety of 360° visibility, and can be opened in flight for the true feel of sport flying. The only real drawbacks are (1) the need to reduce excess solar radiation through the upper canopy; (2) difficult access to the rear seating if the canopy does not move sufficiently far aft; and (3) the fact that everyone might get wet when the canopy is opened in the rain. Overhead glare can be eliminated by ap-

Grumman American

12-6 *Sliding canopy on Grumman American designs.*

plication of heavily tinted plastic liners or the use of darkly tinted glass for the top panels, and proper track design and attention to fuselage lines will permit canopy movement sufficient to open the entire cabin if necessary. Unfortunately, nothing can be done about the weather.

Wing design

While the design of an efficient wing structure for private aircraft was briefly covered during the cost vs. design review (see Figs. 12-2 and 12-3), there are a few reasons why wings are designed as they are that is of general interest. For example, we all know that tapered wings are more difficult to design and build because, among other things, all ribs are of different size. But when compared with a rectangular wing of similar span and area, tapered wings offer the dual advantages of (1) moving the center of lift inboard and (2) increasing the root chord. Moving the center of lift inboard reduces the root-bending attachment load because the moment arm is shortened; this in turn decreases the load that must be taken by the fuselage attachment and carry-through structure and so reduces structural weight. The increased root chord makes the beam deeper at the root, which further reduces attachment loads and structural weight. As another gain, the increased root chord provides more volume for main gear retraction. In addition to their favorable effect on rolling rate as explained in chapter 5, these are more reasons why tapered wings are used on high-performance aircraft. As an esthetic note, they also make a prettier airplane.

The safety advantages of locating all fuel out in the wings has been covered in chapter 11; the structural advantages are presented in Fig. 12-7. This diagram also shows why root-bending loads decrease with a tapered wing as compared with a rectangular planform; because the center of the tapered area is nearer the side of the fuselage, the moment arm will be shorter and the bending moment correspondingly lower as noted.

Note that when the lift force or wing loading is up, the fuel and wing weights act down to provide a reduction in root bending and vertical shear in accordance with

Moment at wing root attachment = M_W = (lift × lift arm) – (wing weight × wing weight arm) – (landing gear weight × l.g. weight arm)

Vertical shear = V_W = lift – wing weight – landing gear weight

If the wing has more than one fuel tank per panel, the inboard tank or tanks should be emptied first if possible; working from the root out in succession provides the maximum reduction in wing bending and shear loads from the remaining outboard fuel. And this load relief acts throughout the flight envelope. When the wing receives a 3 g lift loading from gust or maneuver action, the load-relieving wing structure and fuel weights also increase by the same 3 g inertia factor (acting down to reduce wing bending and shear in the same ratio experienced in steady 1 g flight).

12-7 *The effect of wing structure and fuel weights in reducing wing panel root attachment loads (rectangular wing shown).*

Although the wing weight has already been considered as a concentrated load for simplicity, in actual analysis this weight would be distributed along the span in proportion to wing chord and structural weight variation. Of course, landing gear and external stores fastened to the wing also act to relieve wing bending and vertical shear loads.

Note that, although wing fuel weight actually relieves wing bending in flight, the relieving weight of fuel carried in wing tanks cannot be considered for wing design because it is possible to fly at gross weight with a heavy load in the fuselage and very little fuel in the wings.

Because the way wing loads are developed from airfoil coefficients has been covered in chapters 2 and 6, with the effect of aspect ratio and flaps on wing lift and drag also reviewed in chapter 6, this information is not repeated. But as a final wing-design consideration, every high performance small aircraft should include flaps to improve short-field takeoff and landing performance without unduly penalizing cruise.

For simplicity and effectiveness, the single-slotted flap provides excellent lift at moderate drag as shown in Fig. 6-3. Although the Fowler flap offers the greatest increase in lift, the mounting structure necessary to permit the aft and down extension movement required by this flap is quite complex in comparison with the fixed hinge point slotted flap. As a result, the true wing-area-increasing Fowler flap is seldom used on private aircraft.

Tail, control surface, and flap design

Tail surfaces normally consist of four assemblies: fin, rudder, stabilizer, and elevator. On some aircraft the stabilizer and elevator are combined into one surface, referred to as an all-moving horizontal tail or stabilator.

In either case, the fin and stabilizer or stabilator usually have two beams: one located along the chordwise center of pressure of the surface area, and the other positioned just ahead of the hinge line. Although nothing is particularly exciting or novel about the design of tail and control surfaces, there are a few crucial criteria. Most important is adequate static balance of the movable surfaces to prevent flutter, particularly for higher-performance aircraft.

The amount of balance necessary can vary from none for low-speed aircraft up to 100% for high-speed designs. "No balancing" means the control surface has no weight added ahead of the hinge line to counterbalance the structural weight aft of the hinge line, while 100% balance means a surface is completely statically balanced about its hinge line. There is also need for dynamic balance of the movable surfaces, but this can only be determined by the vibration and flight tests briefly discussed in the following section on flutter.

From a design viewpoint, the ideal way to balance control surfaces is by incorporating a heavy box structure nose section ahead of the hinge axis with very light construction aft of the hinge line. Earlier aircraft accomplished this by use of fabric-covered light alloy frames, but to reduce maintenance this practice has given way to all-metal assemblies. Of course control surfaces also include ailerons, which can be quite crucial in developing wing flutter unless adequately balanced and carefully flight-tested through V_D speeds.

Aerodynamic balance of control surfaces also becomes important at higher speeds or for relatively large control areas. Aerodynamic balancing is accomplished by locating part of the movable area ahead of the hinge line to reduce the surface loads that must be overcome by pilot control forces. Examples of this type of balance are shown in Fig. 7-3, where the area ahead of the hinge line acts to reduce surface forces aft of the hinge line.

Control surfaces provide greater lift per degree of displacement when the gap between the wing or tail structure and the control surface is sealed to prevent airflow between the high- and low-pressure sides of the control surface. Greater lift means a more responsive control system; but because gap sealing involves more assembly parts and work as well as some additional maintenance, sealed surfaces are usually found only on higher-performance and aerobatic aircraft. Excellent background material for the design of aerodynamically balanced and sealed control surfaces is given at the end of this chapter.[6,7]

New aircraft designs might just as well incorporate elevator and rudder trim controls right from the first preliminary studies. Current FAA design specifications require flight demonstration proof that an airplane can be flown by use of the trim controls if any part of the primary control system should fail.[8] While it seems difficult to imagine an in-flight accident that would sever any part of the primary control system without also damaging trim controls, it is possible to use the trim system to provide emergency elevator and rudder control response rates approaching primary control system levels.

Typical methods of light aircraft control surface and flap construction are shown in Fig. 12-8. Some modern fabrication techniques use a beaded surface rather than placing supporting ribs between load points, while some surfaces are bonded together under pressure instead of being riveted. Spotwelding is not commonly used for control surface assembly because this process is subject to fatigue and local corrosion in aluminum alloy structures, as well as being relatively expensive. But regardless of the construction selected, tail surface thickness should not exceed 12% for the reasons noted on page 106 of chapter 7, while the aileron and flap sections will be faired into the wing structure.

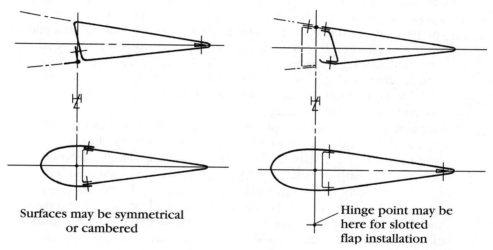

Surfaces may be symmetrical
or cambered

Hinge point may be
here for slotted
flap installation

12-8 *Typical structural design for simple control surface and flap assemblies.*

Those flat sides you might have noticed on control surfaces tend to keep the assembly from oscillating in flight by preventing flow changes due to curvature. To prevent surface flutter and nose-down trim from being induced by a drooping control surface, the simple piano-hinged sections shown in Fig. 12-8 should be at least 80% balanced about the hinge line when used as an elevator. This can be accomplished by a spring or concentrated weight mounted ahead of the hinge line (known as a bobweight), but must be flight-demonstrated to prove the system free from flutter for either design. In addition, if a spring balance is used the system must be free from flutter at V_D with the spring disconnected.

Control surface design can be a bit more complex than it might appear at first. The proof of an acceptable system lies in flutter-free, responsive surfaces providing balanced control forces.

Flutter design and tests

The subject of flutter has filled many books and can easily lead to much computer work, so the causes and possible cures for this undesirable phenomenon that can occur in wing, tail surface, and fuselage structures is all that is covered here.

What is generally referred to as *flutter* can actually be one of three somewhat related but different structural effects: actual flutter, divergence, or control surface reversal. To clarify these by definition:

- *Flutter* is an unstable, self-excited oscillation of definite period set up in an airfoil and its associated structure by aerodynamic, elastic, and inertia effects that have combined to obtain energy from the airstream. Flutter begins suddenly when the crucial speed for a structure has been reached, and it increases in amplitude and intensity with increasing speed. So unless flight speed is immediately reduced, a fluttering structure can disintegrate. High-frequency flutter oscillations induced in heavy aircraft at high speed usually destroys the structure before any corrective action can be taken. Flutter is also more likely to occur at altitude where reduced density atmosphere exerts less damping effect on the wing and tail structures. Tail flutter can also induce fuselage flutter in torsion and bending.

- *Divergence* is the instability of a wing or tail surface in torsion, and can be crucial when the center of lift is ahead of the center of twist (or torsional center) of a structure. For that condition, wing lift tends to increase wing angle of attack, which further increases wing lift, etc. With a torsionally stiff wing, such as the D-spar leading-edge torque box design of Fig. 12-2, at speeds below some crucial "structural divergence" speed the additional increments of twist with airspeed become smaller and equilibrium is obtained without any undesirable effects. In this condition the structure is said to be convergent.

- *Control surface reversal* occurs when the fixed supporting surface is deflected in torsion (twisted) by the control surface in such a manner that control response is opposite to that desired. For example, deflecting an aileron down produces a lift force on the wing structure that will normally be aft of the wing center of twist (or torsional center). If a wing is torsionally weak, it will be twisted leading-edge down, depressing instead of raising the wing panel.

In similar fashion, a torsionally weak stabilizer can be deflected leading-edge down when the elevator is moved down; this could provide quite a bit of activity in the cockpit because the airplane would abruptly nose up instead of trimming nose down as expected.

Rigid structures with minimum play in all hinges, control system components, and trim tabs are essential for a flutter-free airframe design.

Because all flutter-related structural deflections quite obviously result in very undesirable and dangerous flight handling characteristics, extensive ground, analytical, and flight tests are required to demonstrate that an airframe is free of flutter deflections at speeds 10% or more beyond the posted red line limit V_{NE} (never exceed speed = 0.9 V_D per chapter 2 load review).

Ground vibration surveys are first conducted to determine whether any parts of the structure might be excited by engine operating frequencies and so start to flutter in flight. In the process, airframe noise entering the cabin area can be reduced. If you should ever see an airplane undergoing a vibration survey you will realize the level of airframe noise possible from engine excitation.

Tests are usually conducted with low-power electric motors turning remotely driven eccentric weights, and there is no doubt when a section of wing, tail, and fuselage skin or structure becomes crucial and "takes off." The resulting resonant vibration develops considerable noise from low power input; just imagine what noise levels and structural deflections can be developed from 150- to 300-hp engines if they are not properly vibration-isolated, or if the airframe structure is not sufficiently damped.

To satisfy the important flutter-prevention requirements, all areas of crucial vibration must be reinforced or damped in some manner to move their natural vibration frequency outside the excitation range. This is particularly important to assure dynamic balance of control surfaces prior to flight test.

Wing torsional deflection tests are also required by FAR Part 23 design specifications and must satisfy certain empirical criteria intended to eliminate divergence and control surface reversal.[9] If you are interested in flutter, several reports listed in the endnotes provide excellent comprehensive design sources covering flutter prevention, as well as approved ground and flight test procedures.[10]

Because flutter can be crucially affected by control surface balance, there are a few maintenance items that should be observed as part of every pilot's anti-flutter campaign. Mud, snow, slush, or ice accumulation on control surfaces will upset their balance and should be completely removed during preflight inspection. In similar manner ice, dirt, and dust can become lodged inside a surface aft of the hinge line and destroy the existing balance; dirt and dust buildup should be checked during every 25-hr airframe inspection, with the possibility of interior ice checked during every cold-weather preflight.

If flutter should occur from any of the preceding deposits or a local structural failure, reduce power at once and bleed off airspeed to some level about 30% higher than stalling speed. Be extremely careful to retain a comfortable margin of control speed plus coordinated controls with the ball centered, and land as soon as possible. Do not ever attempt to continue flight with a structure ready to flutter if given the opportunity; it will surely occur as soon as crucial flight speed has been reached, and then probably for the last time.

In addition to clarifying just what causes flutter and indicating methods for prevention, this brief review has been included to let you know that every certified airplane has been well tested to prevent any possibility of flutter when operated within the approved weight, balance, and flight envelopes as originally designed and certified.

Composites

Purchase price has such an important effect on commercial sales that increasing amounts of design effort are directed toward development of composite structural materials to reduce airframe cost. Combinations include fiberglass or other synthetic fabrics bonded to various rigid foam materials; graphite, boron, or similar filament fibers in epoxy-type binders; honeycomb core materials with metal or impregnated synthetic fabric facing; chopped fibers, cloth and synthetic resins; etc. See the notes at the end of this chapter for an excellent review of these possible combinations, including production methods and their effect on reducing aircraft cost through automation.[11]

Interestingly, the 1975 Bellanca Aircraft Engineering Skyrocket II shown in Fig. 12-9 has an airframe fabricated from epoxy-prepreg composite material and aluminum honeycomb. In addition to providing a structure that can be assembled by relatively unskilled labor, the resulting synthetic surface is aerodynamically superior to that of aluminum alloy skin and rivet construction—hence the world speed records won by Skyrocket II.

However, structural tests of composite material combinations used by a company for airframe assembly must demonstrate material strength repeatability comparable to sheet metal and extrusion consistency to meet FAA regulations. The FAA requires an ultimate factor of safety of 2 (vs 1.5 for sheet metal) for all structural components fabricated from composite materials.

Bellanca Aircraft Engineering

12-9 *Bellanca Aircraft Engineering Skyrocket II.*

Obviously this imposes a severe empty weight penalty, while the many test samples necessary to demonstrate consistent strength values add up to considerable development cost.

Tooling costs are also very high compared to metal assembly because a plug and mold or molds are required for each component. Some form of vacuum bagging plus a suitable oven capable of providing evenly distributed temperature of at least 250°F is also required to cure simple composite details and assemblies.

Further, my recent experience has shown that it is not possible to produce assemblies for small aircraft comparable to the weight of aluminum alloy components. That is, unless very expensive materials such as Kevlar or carbon are used with a high temperature resin that is cured in an autoclave capable of at least 100 psi pressure and 450°F temperature. Because this type of composite assembly becomes extremely costly to tool and fabricate, it is well beyond the capability of most personal aircraft companies and is restricted to large commercial aircraft and military production.

Regardless of the type of composite construction, the crash and lightning protection provided by a composite structure will be less than that of a metal airframe. However, special light-weight metal and wire sheets can be imbedded in the skin surface to provide a lightning path over the composite structure. Unfortunately, all this adds up to rather expensive and time-consuming tooling and composite assembly, which at present outweighs any possible advantage of somewhat less expensive labor.

Reducing cabin noise

Because the powerplant is the primary source of engine noise, improved methods of engine suspension and isolation are needed that will totally prevent engine vibration from getting into the airframe. Once lightweight skin panels become excited by engine frequency, considerable noise is developed, as mentioned in the flutter discussion. Heavier skins or dampened skin panels will lower the cabin noise level, while dynafocal engine mounts are quite effective in isolating engine vibration during flight.

The need for a well-damped structure also applies to window areas, which require thick acrylic or similar plastic to provide an effective noise barrier. The one remaining weapon against interior noise is the most obvious—soundproofing. And the frequently overlooked firewall area should be a prime candidate for considerable soundproofing protection. If properly applied, modern soundproofing materials will also provide the insulation and panel-deadening (vibration-damping) properties that are most desirable.

Unfortunately, but typical of aircraft development, these improvements are accompanied by increased weight and cost. In return, the resulting airframe should be quieter and the airplane considerably safer for long trips due to reduced fatigue and improved cabin communication. And what more can be expected than a safe structure and quiet airplane. Don't accept less at the present state-of-the-art of aircraft design.

Endnotes

[1]FAR 23.572 (a) (2).

[2]FAA Advisory Circular 43.13-2A, pp. 65-75.

[3]FAA Advisory Circular 23.562-1, "Dynamic Testing of Part 23 Airplane Seat/Restraint Systems and Occupant Protection," June 22, 1989, pg. 26.

[4]Karl Bergey, "The Design of Crashworthy General Aviation Aircraft." *Society of Automotive Engineers Paper 740376*, April 1974.

[5]FAR 23.349 (a) (2).

[6]L. E. Root, "The Effective Use of Aerodynamic Balance on Control Surfaces," *Journal of the Aeronautical Sciences*, April 1945, pp. 149-163.

[7]Hoerner, *Aerodynamic Drag*. Published 1951.

[8]FAR 23.677 (b).

[9]FAR 23.629.

[10](a) J. E. Younger, *Structural Design of Metal Airplanes*, chapter XV, pp. 271-277. McGraw-Hill Book Company, New York, 1935 and subs.

(b) W. S. Diehl, *Engineering Aerodynamics*, chapter 7, pp. 228-234. The Ronald Press Company, New York, 1939 and subs.

(c) ANC-12, *Vibration and Flutter Prevention Handbook*, April 1948 as revised.

(d) "Simplified Flutter Prevention Criteria for Personal Type Aircraft." *Report No. 45*, Airframe and Equipment Engineering Branch, Department of Commerce (now the FAA), as corrected to February 1952.

(e) *Flight Test Report Guide*, pp. 70-73; FAA, May 1969.

[11](a) San Diego Aircraft Engineering, Inc., "Potential Structural Materials and Design Concepts for Light Aircraft." *NASA CR-1285*, March 1969.

(b) A. C. Marshall and J. Brentjes, "Use of Honeycomb and Bonded Structures in Light Aircraft." *Society of Automotive Engineers Paper 730307*, April 1973.

(c) H. D. Neubert and R. W. Kiger, "Modern Composite Aircraft Technology," *Sport Aviation*, July, September, and December 1976.

(d) Brian C. Hoskin and Alan A. Baker, *Composite Materials for Aircraft Structures*. American Institute of Aeronautics and Astronautics, Inc., New York, 1986.

13

Equipment requirements

Although the preceding chapters have been primarily concerned with handling, performance, and structural aspects of aircraft design, these are all rather hidden features inherent in the basic airplane. What seriously affects pilots' comfort and safety, as well as their opinions of a new model, are those items used on every flight—the equipment.

Equipment includes seating and harness, all controls, instruments, radios, and such major systems as airspeed, electrical, hydraulic, vacuum, oxygen, cabin heating, defrosting, and ventilating (which can include air conditioning). As long as the seating and control locations are comfortable and all systems are functioning well, we have a happy crew. But let a radio operate intermittently, the battery spill over, or one omni go out during an IFR hold, and the entire flight goes sour.

The airframe, engine installation, and handling characteristics have not changed one bit, only one or two pilot contact items have failed; but since this experience directly affects the operator, it is remembered as an extreme annoyance, and amplified by thoughts of the installation cost.

Aircraft designers all too frequently concentrate primarily on airframe and powerplant details, throwing in equipment at the last minute to balance the total package. Poor ventilation is a classic indication of this approach, although recent designs have shown some improvement, particularly for windshield defrosting. While it is obviously important to have a sound airframe and powerplant design, all this is of no use if the resulting airplane is inconvenient and uncomfortable.

This chapter offers suggestions permitting owners to correct equipment areas and items that might be found unsatisfactory, while also covering design features essential for reducing maintenance. In this regard, accessibility and reliability are important considerations for every equipment installation. In practice this means if maintenance time and charges are to be kept low, equipment must be accessible; if a piece of equipment is used and depended on for flight safety, it must be accurate and reliable. With these objectives in mind, let us briefly consider principal equipment features and operating systems.

Cabin design

All equipment relates in some way to pilot and passenger accommodation. Because coverage of this subject must be general in nature, I'll review operational priorities to help highlight major differences in basic equipment requirements.

Aircraft intended for the serious business and pleasure markets must have as much elbow room as possible for all occupants, along with well-balanced and friction-free controls, a heated Pitot system, and ample panel space for the instruments and avionics necessary for IFR operation. Adequate ventilation, cabin heat, windshield defrosting, seat belts, and shoulder harnesses are also essential for passenger comfort and safe operation.

Sport and aerobatic aircraft have quite different requirements. A tight cockpit and a strong restraint system are necessary for pilot safety and to provide the level of handling ability required. And well-balanced, low-friction controls are equally important for proper "feel" and control response. If the design is an open cockpit type, nature provides the ventilation. Since IFR operation is not expected of aerobatic aircraft, the instrument panel and avionics can be minimum, although an omni would be useful for cross-country flying or an occasional VOR (very-high-frequency omnirange) approach in unexpected weather.

Agricultural aircraft are pieces of flying farm machinery, but this does not mean Spartan accommodations should be the design standard. Ag pilots can fly in extreme heat for more than 10 hrs a day during seasonal peaks. As a result, elimination of pilot fatigue becomes a crucial consideration, with balanced and highly responsive controls being essential to reduce pilot workload. Modern Ag planes include cockpit air conditioning to eliminate heat fatigue and to more fully exclude spray and dust chemicals, thereby reducing cockpit irritation.

Being cool is not the only requirement for comfortable, long-duration operation. Increased attention is being paid to seat design for both pilot comfort and crash protection. And of course, rugged seat belt and harness restraint systems are especially important for this type of activity. Again, complex avionics and instrument systems are not important, although reliable and accurate stall warning, airspeed, altitude, manifold pressure, rpm, and fuel-level indicators are necessary for safe operation. Since Ag planes infrequently fly cross-country, many are equipped with only a portable communications radio.

To round out the picture with another major design variation, sailplanes have special requirements unique to their type. Powerplant instruments are replaced by variometers to indicate rates of climb or sink in thermals. A snug cockpit with heavy-duty shoulder harness and seat belt is mandatory for higher-performance models to prevent pilot contact with the canopy in rough air, as well as to provide off-field landing protection.

To reduce frontal area drag, the reclining cockpit design is now standard for all higher-performance sailplanes. Complex avionics are not required, although a simple omni would be advisable for distance flying and orientation on hazy days. Since sailplanes are not permitted IFR flight during contests, as per the Soaring Society of America rules, there is no need to include panel space for various gyros and navigating equipment items; but an electric turn and bank could be helpful in hazy day flying.

The surface controls of higher-performance sailplanes in particular must be free of any friction since control forces will normally be extremely light, reducing "pilot feel" to minimum levels during most flight conditions.

From this review, it is apparent that a good flight control system is basic to all aircraft. Such a system is one in which:

- All control forces are evenly balanced and feel similar for elevator, aileron, and rudder operation.
- All controls are sufficiently free of friction to permit the surfaces to return to neutral when released in flight.
- The surfaces provide positive and rapid control response about all three axes.

Balanced control forces depend on required control surface areas, the degree or extent of aerodynamic balance used, and the actual geometry of the individual control systems. By working back through each control system using the applied loads calculated for each set of control surfaces, it is possible to determine the maximum control forces that would be expected in flight at maximum speed.

Naturally, all control forces will be lower at lower speeds, but they will reduce in proportion because the same airflow passes over all surfaces. Using this procedure it is possible to design for balanced control forces, although some adjustments will undoubtedly be desirable following initial flight tests. Ball- and needle-bearing pivots offer low-friction control systems, while the criteria for responsive control surfaces have been well covered in chapters 5 and 7.

All engine and flight controls must be located convenient to the pilot. Flight controls should be designed to clear legs and lap to permit full control motion without restriction, and to provide a clear spot for charts or a flight board. Even though many of us believe a stick offers more positive control, the wheel has gained popularity primarily because it does clean up the cabin; although it is a heavier and more expensive installation than a control stick.

You might be reassured to learn that although actual control surface design loads might be comparatively low, modern FAA practice requires the primary control systems, which are the elevator, aileron, and rudder controls, to be designed for maximum pilot effort. This is known as a *jamming load condition* on the premise that a pilot could exert such forces if a surface became locked or jammed in a fixed position. The pilot limit load forces are:

Aileron control
   stick   67 lb
   wheel   50 lb (applied at the rim of the wheel on each side and in

opposite directions)

Elevator control
    stick   167 lb
    wheel   200 lb
Rudder control
    per pedal   200 lb

These forces must be applied left and right, push and pull on the stick or wheel, based on the load conditions listed in FAR 23.397.[1] All values must be multiplied by a factor of 1.50 to obtain ultimate loads for detailed design. Since control system strength must be demonstrated by static load test for each newly certified design, control systems rarely cause any problems other than those of routine service maintenance.[2] To simplify the required 100-hr and annual inspections, access handholes or hinged panels should be provided at all major control system pivot, pulley, and adjustment points.

Because visibility is something impossible to have in excess, an unlimited number of different arrangements have been tried in attempts to improve visibility from the pilot's seat. Starting with the periscope used on Lindbergh's Spirit of St. Louis of the late 1920s and Hawk's retractable cockpit-enclosure-equipped Time Flies of the late 1930s, the ultimate in visibility can be found on some current homebuilt designs.

One of these homebuilt designs with unlimited visibility is the Breezy, where the pilot sits right out in the open air in the best Glenn Curtiss tradition. Somewhere in between lies the desirable balance of at least 240° visibility around the horizon (in azimuth) and 11 to 13° down over the nose. The former is important for visually sweeping the area before starting a turn to either side, while the downward visibility is important during taxi, climb, and approach phases of flight.

The 11° downward vision angle is also quite useful in another way. From an altitude of 3000 ft above ground level, this angle permits sighting the ground almost exactly 3 mi ahead of the airplane. So when the ground disappears in haze, fog, or rain, you had better turn around—you have automatically been advised that conditions ahead are below VFR (Visual Flight Rules) minimums, and with sufficient warning to permit a safe 180° turn while still VFR. The downward and forward vision angle can be readily checked from your normal seat position by leveling the plane to flight attitude, sighting the nearest ground point, and then computing the downward angle based on eye height above ground per Fig. 13-1. Any time taken to determine this in-flight visibility indicator is well spent, in terms of flight safety.

Since cabin visibility means safety in many ways, this becomes an important design consideration. Cessna has provided exceptional flight visibility for a series of single-engine aircraft in which the upper cabin offers an almost 360° sweep as shown in Fig. 1-3, and the vertically adjustable pilot's seat permits satisfactory downward vision over an otherwise high instrument panel.

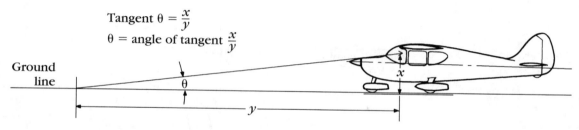

Tangent $\theta = \dfrac{x}{y}$

θ = angle of tangent $\dfrac{x}{y}$

Ground line

13-1 *Determining the downward vision angle θ.*

Pilot seats mounted on tracks that incline up as the seat comes forward should be lighter and much cheaper than the present vertically adjustable ones, and so deserve further consideration. As another design approach, military seats move vertically on guides (as necessary for ejection) while the pedals adjust fore and aft to accommodate different leg positions. Having both pedals and seats adjustable would be very nice for changing position on long flights, and probably is the ultimate arrangement. But because these features increase cost and weight, only movable seats are normally used in small airplanes; although adjustable seats and pedals are quite commonly included on high-performance sailplanes.

Figure 13-2 shows a typical, comfortable modern cabin and control system arrangement. Any additional rows of seats should have at least a 32-in clearance from those in front, and preferably 36 in. If you are interested in developing cabin and cockpit accommodations, the 1/16 scale person and seat arrangement in Fig. 13-3 is useful for preliminary layout and study.

All seats must be designed to carry at least a 970-lb vertical load, with aerobatic seats stressed for 1710 lbs of passenger and parachute.[3] In addition, each seat and supporting structure must be tested to withstand crash loadings of 9 g forward and 1.5 g to either side, plus dynamic loading as discussed in chapter 12.

Seat belt load and installation requirements are similar to those specified for seat design. Seat belts and harnesses must be tested to FAA Technical Standard Order (TSO-C22) and approved by the FAA for aircraft use. Over the years, a great number of materials and release system improvements have gone into seat belt development; if you have ever spun or been upside down in an open cockpit airplane, you appreciate just how comforting and important a strong lap belt can be.

Today's FAA design requirements place equal emphasis on seat belt and shoulder harness installations for crash protection. World War II clearly showed a reduction of about 50% in crash fatalities when harnesses were used, so always use yours. While no one expects to crash, there is precious little time to strap on a harness when weather or other problems dictate a rough field landing. For those of you interested in harness requirements, attachment loads were noted in the fuselage design section of chapter 12, with installation details covered in chapter 12 endnote.[2]

13-2 *A typical cabin arrangement of seating and controls.*

13-3 *A ¹⁄₁₆ scale man, seat, floor, and pedals diagram.*

Thanks to the design requirements of current FAA specifications, when well-strapped and harnessed in place you are able to withstand anything the fuselage can withstand, and many tests have been performed to prove this is so. As becomes increasingly evident, with all the test work required for certification it is no wonder development costs climb faster than the airplane ever will.

Instruments, instrument location, and lighting have probably varied over the years more than any other equipment item. Of course, much of this change resulted from technical advances in instrument quality and reliability. But an equal amount of variation in type and location occurred through individual preference, with every pilot having his or her own idea of the best possible panel layout.

As aircraft came into "mass" production and rental ownership developed, flight safety was frequently related to the speed with which pilots could locate specific panel instruments in an emergency. Following a brief pattern checkout they might be flying solo in an airplane new to them, with instruments not at all similarly located to those of the plane flown for the past 30 or 40 hrs. And so corrective action could be slow when needed, and frequently too late.

To improve this situation, a standard location of the basic IFR flight instruments is now part of the certification requirement for aircraft heavier than 6000-lbs gross weight. This same arrangement is applied by the FAA to all aircraft equipped with attitude and directional gyros, resulting in the familiar T-panel layout in Fig. 13-4.

13-4 *Instruments arranged in the T panel layout of FAR 23.1321.*

To assist in reducing familiarization time between different airplanes, whether homebuilt or production types, this instrument arrangement should be followed for all models equipped with gyro instruments. Note that the attitude and directional gyros are specified as vacuum or pressure driven, while the turn and bank (T & B) is electric. This practice eliminates a second air-regulating valve for the T & B while also providing dual-lateral attitude indication, a backup system to prevent those crucial spiral dives should an electrical or vacuum system fail. If the vacuum system goes out, it is possible for IFR-trained pilots to maintain a level and safe attitude by scanning the (electric) T & B, airspeed, and ROC instruments.

Pilots should frequently check their proficiency in maintaining level flight with these three instruments, but under flight instructor supervision. With practice it is possible to maintain level flight and altitude with just the T & B plus airspeed operating, so the basic IFR instrument cluster in Fig. 13-4 provides overlapping information when all systems are operating.

The single most important instrument system is the Pitot-static (or airspeed) installation. This system is basic to aircraft operation because it provides information for the altimeter and ROC instruments in addition to indicating flight speed. For FAA certification, airspeed indication must be accurate to within 3% or 5 knots, whichever is greater, of true airspeed. Airspeed calibration accuracy is thoroughly checked as part of the certification flight test program, with corresponding airspeed conformity checks performed as part of the flight test procedure for every production airplane. So if your Pitot-static system is kept free of bugs, mud, ice, and water, it will indicate flight speed quite accurately.

There are four ways to ensure reasonably accurate Pitot-static system operation in addition to the normal preflight inspection. This number of backup checks alone indicates the importance assigned to the primary airspeed system. These four checks are:

1 Water drain traps that are bled to remove any condensation or rain that has entered the pressure lines and could result in false readings.

2 An electrically heated Pitot-static (airspeed) head to prevent ice accumulation.

3 An alternate static source that can be opened if the primary static pickup is blocked.

4 A variation of the alternate static, which is to carefully break the ROC face to permit emergency static pressure to enter the system for altitude indication. ROC readings must then be disregarded because this instrument will be working in reverse; use the altimeter as a climb reference if necessary.

The Pitot and static line water drains should be accessible in flight, if possible, just in case trouble is experienced during flight through a severe rain shower. These quick drains must be located at the lowest point in the airspeed lines to permit easy inspection at least every 25 flight hrs, or during preflight whenever the airplane has been tied down in blowing rain.

An electrically heated Pitot-static head is highly recommended for all aircraft because it not only permits removal of unexpected ice accumulation from the airspeed pickup, but heat can help remove water as well. If only one option is possible, the electrically heated airspeed head is preferable to an alternate static inlet because there is no way to open an iced-over Pitot head without heat, while altitude indication can be obtained by breaking into the ROC case as previously mentioned. Or stated another way: when both the Pitot and static pickups are thoroughly out of action due to ice accumulation, the alternate static will be of no use for airspeed indication because there is no dynamic (Pitot) pressure available. Of course, an alternate static will provide acceptably accurate altitude and ROC indication and we will find many modern aircraft equipped with both a heated airspeed head and an alternate static source.

During night operation, lighted instruments must not create confusing reflections all over the windshield. Individual post lights mounted through one of the instrument attachment screw holes provide excellent lighting when all panel lights are wired into a single dimming switch. Any instrument reflections can be contained by mounting a padded shield off the instrument panel top deck, being careful not to block instruments or switches from pilot view.

As a final instrument note, an audible stall warning indicator is practically mandatory for all newly certified aircraft. Safe Flight Instrument Corporation is the leading supplier of a small wing-mounted pickup used on all modern United States airplanes. This installation satisfies FAR Part 23.207 requirements by providing both a red warning light and buzzing signals throughout an adjustable speed range prior to full stall.

If you have instrument work to be done, necessary design and operational requirements for acceptable airspeed indicating systems are given in FAR 91.33, while practical instrument installation and maintenance suggestions have been gathered in AC 43.13—many of which can be readily performed by private owners.[4, 5]

While opening a door on a closed cabin is about the only step the average owner can take to reduce stifling ground heat on a sticky midsummer day, air-conditioning options are now available for some modern aircraft. Whether or not this solution is viable due to weight and cost increases, depends on personal preference and home base location. Certainly our northern areas do not warrant air conditioning, but this equipment must be a great source of relief during southern summers.

Whether air conditioned or not, closed cabin areas must be provided with reliable sources of cold and hot air during flight. Air delivery should be sufficiently quiet not to interfere with radio communications or become a source of fatigue, with outlet flow broadly disbursed and not ejected as if from a fire hose on whoever is nearest each opening. Heat is normally added to fresh outside air by passing through a transfer muff carried as part of the engine exhaust system (Fig. 11-3), and can be selectively mixed with cold air to provide the desired degree of cabin heating.

One ventilation feature often neglected is the frequent need for windshield interior defrosting prior to takeoff. Because there is no moving air in most ventilating systems until airborne, there can be no defrosting without an auxiliary blower—and this is a badly needed item of equipment. Because windshield condensation is common during summer and fall operation in mountainous terrain and on the water, and a regular pre-takeoff experience in wintertime, pilots need to be able to select heated defrosting air during ground operation. To be adequate, ventilation systems should include a windshield defrosting blower as standard equipment.

Electrical systems

Aircraft electrical systems are usually taken for granted by owners and pilots, in much the same manner as for an automobile; get in, turn on the key, and start off. This might be sufficient for VFR sport flying where the most important job an electrical system performs is reliably starting the engine. And even that assistance is not expected on some of the older hand-cranked classic models.

But for cross-country flying where reliable communications, navigation, and various forms of lighting are necessary, a dependable electrical system is essential. For our purposes, this means the basic wiring and any subsequent additions must be performed in accordance with acceptable standards of wire size for the load carried and length of run; with proper switch, fuse, and circuit breaker size; and with reliable electrical accessories, such as rheostats, Pitot heat, navigation lights, panel lights, indicator lights, rotating beacon or strobe lights, and landing lights.

Adding in the radios and related navigating equipment results in a total load of many amps when all items are operating during night IFR conditions. To make sure everything keeps running at maximum demand, FAA certification requires an electrical load analysis proving sufficient alternator or generator capacity is available to satisfy peak amperage. Slight overload can be supplied from the battery for short periods of time—for example, turning on landing lights during final—but load requirements should preferably not exceed 80% of alternator or generator capacity at the maximum continuous running load.

A typical electrical load analysis is simply compiled from equipment item specifications and would follow this format (for a 12-V system):

Item	Maximum amperage
Fuel quantity indicators	0.90
Oil temperature gauge	0.30
Oil pressure gauge	0.30
Fuel pressure gauge	0.30
Ammeter	0.30
Auxiliary electric fuel pump	0.70
Landing gear position indicators (2 lamps)	0.24
Heated Pitot-static head	12.00

Item	Maximum amperage
Dome lights (**)	1.00
Panel lighting—lamps and rheostat	1.04
Internal lighting in compass, T & B, radios	3.00
Navigation lights—wing tips and tail	4.04
Strobe lights (2)	2.40
Landing lights (**)	6.00
T & B (electric)	0.75
Stall warning system	0.12
IFR radio equipment (dual nav., dual com., glide slope, ADF, transponder, DME, Loran or GPS, and audio panel)	15.00
Load with all items operating	48.39 amps
10% overload safety margin	4.84 amps
Total maximum condition	53.23 amps

Items marked (**) are not required for normal cruising flight (long period) operation and so can be discounted. We then have:

Load with (**) items turned off (48.39 - 7.00)	41.39 amps
10% overload safety margin	4.14 amps
Total amps for system design	45.53 amps

To supply this system demand at 80% output, the alternator or generator must be rated at

$$\frac{45.53}{0.80} = 56.91 \text{ amps capacity}$$

Judging from these calculations, a 60-amp, 12-V alternator or generator would be adequate. Similar load analysis will determine the electrical source capacity required for new or modified systems.

As military aircraft grew larger and more complex, 24-V systems became standard, principally to reduce power losses throughout the system. Most small aircraft are standardized at 12 V, now common on automobiles as well, and so recommended wire size, circuit breaker or fuse ratings, and switch charts have been included in Table 13-1 to cover 12-V electrical systems.

The wire run vs. size data can also be safely applied to 24-V systems at half the indicated amperages because the line losses would be lower than for 12-V runs of comparable length. This data should enable any owner to plan wiring changes that will not overheat, although the system load must be rechecked if additional electrical equipment is to be installed and operated during flight.

Any owner feeling up to the task can make many types of wiring changes under the supervision of approved mechanics, and at a considerable saving over repair shop costs. Normal aircraft wire should be of MIL-W-16878 or MIL-W-5086 quality, with the former being superior for higher temperatures and salt-water exposure. Magneto wire should be shielded to meet specification MIL-W-76B or better, with generator and mag filters installed to reduce radio interference noise.

Table 13-1(a)　Electrical system design data: wire length vs. load, circuit protection, and switch selection ((a) and (b) courtesy of Boating Industry Association; (c) and (d) courtesy of AC43.13-1A, page 177)

Wire size	20	18	16	14	12	10	8	6	4
Stranding	7 × 28	16 × 30	19 × 29	19 × 27	19 × 25	19 × 23	19 × 21	49 × 23	49 × 21
Circular mil area	1094	1568	2340	3777	5947	9443	15105	24354	38430
Circuit current, amps	MAXIMUM LENGTH OF CONDUCTOR IN FEET FROM POWER SOURCE TO LOAD AND RETURN								
1.	36.4	52.3	78.0						
2.	18.2	26.1	39.0	63.0	99.0				
3.	12.2	17.4	26.0	42.0	66.0				
*4.	9.1	13.1	19.5	31.5	49.5	78.8			
5.	7.3	10.4	15.6	25.2	39.6	63.0			
*6.	6.1	8.7	13.0	21.0	33.0	52.5	83.8		
7.	5.2	7.4	11.1	18.0	28.2	45.0	72.0		
8.		6.5	9.8	15.8	24.8	39.4	63.0		
*9.		5.8	8.6	14.0	22.0	35.0	56.0	90.0	
10.		5.2	7.8	12.6	19.8	31.5	50.4	81.3	
*15.			5.2	8.4	13.2	21.0	33.6	54.1	85.5
*20.				6.3	9.9	15.8	25.1	40.6	63.4
*30.					6.6	10.5	16.8	27.1	42.7
40.						7.9	12.6	20.2	32.0
50.						6.3	10.1	16.2	25.6
55.							9.2	14.7	23.3
60.								13.5	21.4
75.								10.8	17.1
90.									14.2

Table 13-1(b)
Circuit breaker
and fuse loadings

Amperage	
Normal rating	Maximum continuous load
4	3.6
6	5.4
9	8.1
15	12.6
20	17.0
30	24.0

Table 13-1(c)
Wire and circuit protector chart

Wire AN gauge copper	Circuit breaker amp	Fuse amp
22	5	5
20	7.5	5
18	10	10
16	15	10
14	20	15
12	25 (30)*	20
10	35 (40)*	30
8	50	50
6	80	70
4	100	70
2	125	100
1		150
0		150

*Figures in parentheses may be substituted where protectors of the indicated rating are not available.

Table 13-1(d)   Switch derating factors

Nominal system voltage	Type of load	Derating factor
24 Vdc	Lamp	8
24 Vdc	Inductive (relay-solenoid)	4
24 Vdc	Resistive (heater)	2
24 Vdc	Motor	3
12 Vdc	Lamp	5
12 Vdc	Inductive (relay-solenoid)	2
12 Vdc	Resistive (heater)	1
12 Vdc	Motor	2

Notes:

1. To find the nominal rating of a switch to operate a given device, multiply the continuous load current required by the device by the derating factor corresponding to the voltage and type of load.

2. To find the continuous load current that a switch of a given nominal rating will handle efficiently, divide the switch nominal rating by the derating factor corresponding to the voltage and type of load.

While battery location might depend on balance requirements because every battery represents a fairly substantial mass of lead ballast, it is important that installation be accessible for frequent inspection and service. Every battery, whether self-contained and vented or requiring a battery box, must be vented overboard at a location that will neither siphon the battery nor spill on the exterior (and ruin the finish).

As a structural protection, the area immediately around and below a battery should be coated with a bitumastic-type compound resistant to battery acid. Battery size will depend on starter and engine requirements, but for colder climates a 37 amp-hr capacity should be the minimum for a 12-V system starting 150 to 200 hp.

For electrical system design requirements and useful maintenance suggestions, consult the endnotes 6 and 7 at the end of the chapter.

Radio systems

The FAA design regulations say little about radio installations and they don't need to. Pilot opinion develops quite strongly and rapidly whenever radio equipment operates intermittently or fails completely. There is nothing more disturbing than an ADF (automatic direction finder) that won't pick up a strong beacon 6 mi away today, but unerringly homes in on a weak one tomorrow.

It is common practice to include all communicating, navigating, and antenna items under the general terminology of radio (or avionics) equipment. Modern solid-state radios are available at all levels of complexity and reliability, with prices increasing right along with the degree of sophistication. While model selection is a matter of personal opinion and operating requirements, it does not seem necessary to purchase TSOd equipment for normal use, including IFR operation, if dual nav/com units are installed.

In view of the great amount of material available concerning radio and antenna installations, with a few of the more useful ones listed in the endnotes, I'll only cover a few pertinent items in this chapter. It might help to know it is possible to drive two omnis plus a glide slope receiver from one V or "balanced loop" antenna mounted on the fin or fuselage. The antenna lead-in goes directly into a small "black box" known as a triplexer, which divides the signal received into components required by each omni plus the glide slope receiver.[8]

If only two omnis are used, or one omni with glide slope, the single antenna lead-in would go into a diplexer that converts the incoming signal into two parts for omni and glide slope presentation. The point being that even when three items of navigation or instrument landing system equipment are used, all can be fed from one omni antenna by use of couplers, as the diplexer and triplexer are known. So two or three receiving antennas are not necessary, although they will be required if couplers are not used.

Whether installing new or add-on equipment, a small ball-ended transponder-type antenna can be used quite successfully as a DME antenna, rather than the more familiar and expensive blade type. When dual communications sets are installed, two com antennas are mandatory; for best reception these should be mounted more than 2 ft apart and not too close to the ADF sense antenna. Figure 13-5 shows a well-planned, modern antenna system suitable for small-aircraft IFR avionics.

13-5 *Modern antenna installation on the Commander.*

Back in the cockpit, the DME package must not be panel-mounted too close to the ADF receiver because it can emit energy that interferes with ADF reception causing pointing errors. Otherwise the avionics equipment seems to be quite compatible, provided all items are properly wired.

When more than one nav/com system is installed along with an ADF and transponder, DME, marker beacon or Loran/GPS, a single control panel should be used to bring everything together and also permit instant switching between preselected frequencies. This convenience is most useful when departing a busy airport, communicating en route, or making an IFR approach.

Communication and navigating equipment that operates unreliably and cannot be depended on for accuracy is worse than no equipment at all. In view of this it is necessary that all radio items used for serious IFR operation be properly installed with adequate cooling, equipped with recommended antenna and lead-in accessories, and maintained in satisfactory condition.

Vaccum/pressure systems

Some type of power source is required to operate gyro-driven instruments. To provide instrument redundancy, current practice prefers the T & B indicator to be electrically driven, while the attitude and directional gyros are powered by an engine-mounted vacuum or pressure system pump.

The pressure system is preferred for aircraft operating at higher altitudes or with pressurized cabins, while most of our lower-performance general aviation models are presently powered by vacuum systems. Most new installations will probably be of the pressure type because this system provides greater mass air-flow at altitude, resulting in better pump cooling, higher efficiency, and im-proved gyro life. It also offers an installation with less plumbing, as shown in Fig. 13-6.

13-6 *Vacuum and pressure systems suitable for gyro operation. (a) Schematic layout for a typical vacuum system. (b) Schematic layout for a typical pressure system. (Based on Airborne Manufacturing recommendations.)*

While the owner's manual or manufacturer's data should be checked for rec-ommended operating limits, most gyros are designed for 4 to 5 in. of mercury differential pressure or vacuum, with 4 in. considered about the optimum for at-titude and directional gyros.

Anyone interested in installing or relocating gyro equipment will find basic design requirements in FAR 23.1331, including the need for two separate instru-ment power sources for multiengine aircraft. The note at the end of the chapter offers useful installation data and sample calculations for adding gyros, while Fig. 13-7 can be used as a plumbing guide.[9] Incidentally, the average total air consumption of 4 cu ft/min for the attitude and directional gyros has been noted on the chart in Fig. 13-7.

13-7 *Line pressure drop chart for smooth tubing used for vacuum and pressure pump systems. (After AC43.13-2, Fig. 11.1.)*

Hydraulic systems

Although most fixed-gear, small aircraft do not have complex hydraulic pump, selector valve, and piston actuated hydraulic retraction systems, they definitely do have one hydraulic system—the brake system. And as required for all hydraulic installations, brake systems must be static tested at 1.5 times the maximum operating pressure per FAR 23.1435.

Since hydraulic system design, installation, and maintenance have been thoroughly reviewed in many standard reference books, in addition to the coverage listed in the endnotes, detailed discussion is not necessary here.[10] To reduce system weight, cost, and maintenance, designers should consider a double-acting hydraulic hand pump and accumulator for the source of hydraulic power.

Pumping up the accumulator while en route would build up a reservoir of pressurized fluid sufficient to lower and positively lock the landing gear when entering the pattern. Similarly, hydraulic pressure could be stored up before starting the engine, permitting instant gear retraction after takeoff when the selector valve is moved to "wheels up" position. After all, the operating cylinders respond to pressure built up by muscle power just as fast as they do to pump power, and the simpler system results in fewer hydraulic headaches plus lower maintenance costs.

Oxygen requirements

Although few of us now fly at altitudes requiring oxygen assistance, this will increase with the growing use of turbocharged powerplants. Present rules require the flight crew to use supplemental oxygen if remaining at cabin altitudes between 12,500 and 14,000 ft mean sea level (MSL) for periods exceeding 30 min. Operation at cabin levels greater than 14,000 ft MSL is prohibited unless at least the flight crew uses oxygen at all times; and every occupant must be provided with supplemental oxygen at all cabin pressure altitudes greater than 15,000 ft MSL (FAR 91.211).

Certification of supplemental oxygen systems can be obtained with portable equipment, and this type of installation should be adequate for most small-aircraft requirements. Approved oxygen systems are available from a number of reputable manufacturers who offer detailed information for their different systems. If oxygen equipment is desired or planned for the near future, the design requirements plus the installation and service recommendations given in endnotes 11 and 12 should be quite helpful. Again, in view of the extensive literature covering oxygen systems, it is not considered necessary to provide more detailed information in this chapter.

IFR equipment

Equipment and installation reliability affect the level of IFR safety. Depending on the amount and degree of IFR operation normally undertaken, autopilot and deicing systems might be required in addition to the desirable dual nav/com radio installations plus ADF, transponder, Loran or GPS, and DME equipment. To be useful, the autopilot must be maintained in adjustment, the gyros must operate consistently, and all avionics equipment must perform with repeated accuracy; anything less represents a decrease in flight safety.

Such demanding requirements show the equipment complexity necessary for sophisticated flight operation. Of course, initial purchase and routine maintenance costs also increase with equipment quantity. For an airplane to be successful, equipment items cannot be left until last; their type, location, and accessibility must be considered along with the airframe as the design progresses. I trust this chapter has indicated how and why.

Endnotes

[1] FAR 23.397, 23.399, 23.405.
[2] FAR 23.395 and 23.651(a).
[3] FAR 23.561, 23.785, and 23.1413.
[4] FAR 91.33, 23.1321, 23.1323, and 23.1325.
[5] (a) AC 43.13-1A, pp. 303-310.
 (b) AC 43.13-2A, pp. 85-88.
[6] FAR 23.1351 23.1401.
[7] (a) AC 43.13-1A, pp. 173-212.
 (b) AC 43.13-2A, pp. 77-83.
 (c) William J. Kendall, "Your Alternator or Generating System," *Business and Commercial Aviation*, May 1976, p. 60.
[8] (a) AC 43.13-1A, pp. 291-301.
 (b) AC 43.13-2A, pp. 5-24.
 (c) John Ferrara, *Every Pilot's Guide to Aviation Electronics*. El-jac Publishing Co., 133 Glendale Drive, Trenton, N.J. 08618.
 (d) Richard N. Aarons, "Antennas? Who Cares," *Business and Commercial Aviation*, June 1974, p. 52.
 (e) Richard N. Aarons, "Planning an Add-on Radio Package," *Business and Commercial Aviation*, January 1976, p. 74.
 (f) Richard G. Lawrence, "Catering to Your Black Boxes," *The AOPA Pilot*, January 1976, p. 44.
[9] AC 43.13-2A, pp. 85-88.
[10] AC 43.13-1A, pp. 165-172.
[11] FAR 23.1441-23.1449.
[12] (a) AC 43.13-1A, pp. 154-158.
 (b) AC 43.13-2A, pp. 43-47.
 (c) Hamilton Westfall, "Oxygen," *Aero Magazine*, April 1976, p. 16.
 (d) D. H. Gollings, "Oxygen and Its Effects on Flight," *Air Facts*, May 1975, p. 6.

14

Landing gear design

During the early 1940s, aircraft design arguments over the advantages of conventional vs. tricycle landing gear were only rivaled by the frequent arm waving and shouting considered necessary to protect buried liquid-cooled engines from those new, big, round air-cooled units. In each case, the new development seems to have been the victor and sole survivor—until recently. Today, increasing numbers of aircraft are fitted with conventional landing gear for special-purpose applications, such as Ag planes and acrobatic types, while time has replaced air-cooled radials with jet engines for high-performance operation.

Types of landing gear

I have never seen a classification of the different types of landing gear built and flown over the years, although the many variations merit some recognition. Considering only the most frequently used and, therefore, probably the most successful, are:

Conventional landing gear

Conventional landing gear, also known as a *tailwheel* or *tail dragger*. The main landing gear wheels are located on each side of the centerline ahead of the center of gravity, with a steerable tailwheel located aft near the rudder (originally a fixed-tail skid was used and can still be seen on some antique aircraft). This type of landing gear was noted for its ability to land "tail first," i.e., ground loop. Why this can happen will be explained a few pages further along.

Tricycle landing gear

Tricycle landing gear, known as the "nosewheel" type. It was orignally developed by Glenn Curtiss back in 1908. The main wheels are located on each side of centerline behind the center of gravity, with a free-swivel or steerable nosewheel mounted on centerline forward (just behind the propeller on most small landplanes). Usually a streamlined skid is fitted aft near the sternpost to prevent scraping the fuselage or tail surfaces during an extreme tail-down landing. This configuration is noted for ease-of-ground handling and will not ground loop unless steered into a skid.

Four-wheel landing gear

Four-wheel landing gear with two wheels mounted on each side of the airplane centerline in tandem fashion. The main wheels are normally located behind the center of gravity, with smaller wheels at the forward position customarily occupied by a single (tricycle gear) nosewheel. This type of wheel arrangement is frequently used with amphibious floats, in which case the small wheel forward is mounted on the bow of the float. This is a very stable system if the forward wheels are free-swiveling.

Tandem-wheel landing gear

Tandem-wheel landing gear with two wheels mounted one behind the other on the airplane centerline. The larger wheel is usually located just behind the center of gravity, with a smaller wheel forward in a nosewheel position. This arrangement has been carried to some sort of ultimate in the B-52 bomber design, which has dual sets of tandem wheels retracting into the fuselage ahead and behind the bomb bay, plus outrigger wheels retracting into each outboard wing panel. With tandem gear it is difficult to taxi in a crosswind, even with a steerable front wheel.

The single-wheel gear

The single-wheel gear, usually mounted just forward or aft of the center of gravity with a landing skid to protect the fuselage. This type of gear is found on most sailplanes, with the wheel being manually retractable on higher-performance designs.

Skid gear

Skid gear, used by the Wright brothers for their early designs and then again through World War I, was frequently incorporated into wheel landing gear systems for nose-over propeller protection. Skid gear is used on modern helicopters, and has been proposed for recoverable space vehicles.

Snow skis and floats can be used with any of the preceding systems. It is interesting to note that all of these landing gear arrangements, and many others, were designed and flown before World War I. Glenn Curtiss's 1908 June Bug was equipped with tricycle gear, a feature carried on all of his designs through 1913 and adopted by the Wright brothers, Breguet, and Cody during 1910.

Dual-wheel conventional-type gear was also very popular with early designers. Practically all aircraft types used this gear at some time and with minor variations. Because available bicycle wheels could not support aircraft weights at high speeds on rough ground, two bicycle wheels were frequently mounted side by side on the same axle, usually with a protective skid between to save the propeller and engine during nose-over accidents.

Four-wheel gear was developed by AVRO in 1907 and was subsequently used by Sopwith, Bristol, M. Farman, H. Farman, and Cody among others. Tandem-wheel gear was first introduced on the 1907–1910 series of REP aircraft and has reappeared over the years for special applications, but remains the least popular of surviving landing gear configurations.

So much for early landing gear history; let us now examine the relative merits of conventional and tricycle landing gear from the design and operational standpoints.

Conventional gear

Conventional landing gear design normally places the main wheels on a line running about 16.5° ahead of the center of gravity, with the airplane in a level attitude, as shown in Fig. 14-1. This position is based on a study of 10 aircraft, including twin-engine amphibians. For aircraft having a high thrust line, it will be necessary to increase the angle by 4 or 5° to prevent nose-over during engine runup and rough-field operation. However, the further forward the main wheels are placed, the larger the tailwheel loading. And the heavier the tail load, the greater the tendency to ground-loop, as can be noted from Fig. 14-2.

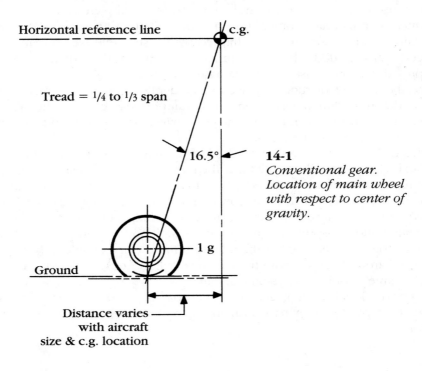

Horizontal reference line — c.g.

Tread = 1/4 to 1/3 span

16.5°

1 g

Ground

Distance varies with aircraft size & c.g. location

14-1
Conventional gear. Location of main wheel with respect to center of gravity.

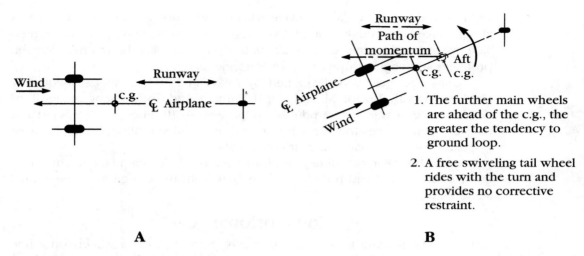

1. The further main wheels are ahead of the c.g., the greater the tendency to ground loop.

2. A free swiveling tail wheel rides with the turn and provides no corrective restraint.

A   B

14-2 *Stability characteristics of conventional landing gear. (a) Conventional gear tracking down runway. (b) The ground loop starts. If the momentum vector gets outside the tread, a ground loop is underway. Note that the aft center of gravity requires a smaller skid angle to start the windup. Tracking correction must be made immediately.*

A free-swiveling tailwheel is likely to ground-loop during crosswind operation because it is free to turn with the tail end of the airplane, and so offers no resistance to rotational momentum once the tail starts to swing. Fortunately, most tailwheel-type aircraft are equipped with a steerable wheel that remains engaged until the rudder has been turned about 30° off the airplane centerline. This provides lateral resistance to sidewise motion of the tail. Unfortunately, 30° can be the amount of rudder deflection suddenly applied to correct a severe skid condition. And then, just when tail restraint is most required, the tail wheel releases into free-swivel and away you go upwind toward the woods or runway lights as shown by Fig. 14-2b.

This embarrassment can be reduced, if not eliminated, by use of a locking tailwheel. This system physically locks the tailwheel to the fuselage or tail spring structure by a cockpit-controlled pin actuated during the landing approach check. Before World War II, when conventional gear planes were practically the only type built, all carrier-landing aircraft used locking tailwheels, as did most of our military and transport designs. But because it cost something to install and was one more complication for the pilot to unravel (for example, it is very difficult to taxi around corners with the tailwheel still locked), the locking tailwheel never became popular on personal aircraft.

If conventional landing gear can be a bit much to handle in crosswind conditions, why does it persist? Principally because this arrangement is excellent for

- Rough-field and short-field operation. The tail can be raised clear of terrain early in the takeoff run and trim attitude set for maximum wing lift, as discussed in chapter 6. If it were possible to raise a tricycle gear nosewheel clear of the ground at 20 mph, so much elevator deflection would be required that the accompanying drag would prolong or prevent takeoff (see Fig. 6-1).
- Amphibian operation. Without a nosewheel it is possible to approach a ramp in cross current and/or crosswind conditions without first grounding on the nosewheel and pivoting about to lie across the ramp. With conventional gear both main wheels touch the ramp at the same time, permitting direct taxi up and out onto the parking area. And when beaching, applying power lifts a tailwheel clear but buries a nosewheel in the sand.

These features are accompanied by lighter installed weight, cheaper manufacturing cost, and reduced maintenance compared to a tricycle gear. As a result, the conventional gear is particularly well suited to agricultural and sport aircraft and is extensively used on these types today.

Those of us who fly for pleasure or occasional business purposes should be alert when taking off or landing crosswind with conventional gear. But if you keep ahead of the airplane, practice will develop confidence. Because of main gear location, tailwheel load, and tread some conventional gear aircraft handle easier than others; for example, I know my PT-19 handled much better than any PT-17 or Luscombe 8 ever did during landing runout. And grass airfields were much easier to operate from than paved runways.

To relieve the crosswind landing problem, development of automatic or semiautomatic runway alignment landing gear flourished during the mid- and late 1940s. Generally known as *crosswind* or *self-castering* gear, various designs were developed by J. Geisse, Goodyear Aircraft, and B. W. King among others.

The principle of operation is shown in Fig. 14-3; the main wheels were arranged to caster and trail along the path of momentum straight down the runway. They occasionally would also caster without warning when taxiing around corners, were heavier and more expensive than the plain wheel and axle combination, and required more service adjustment and attention. These disadvantages, combined with the arrival of the new generation of postwar tricycle-gear aircraft, precluded full development and acceptance of the crosswind gear concept.

Incidentally, Louis Bleriot first incorporated this type of gear in his English Channel crossing airplane back in 1909. A version of the old castering main wheel is still with us on the C-5 military transport. This huge airplane can steer its main gear up to 18° from the fuselage heading to land and run out in a crabbed attitude. Heaven help us all if the pilot should forget just before touchdown and kick out the crab with the main wheels cocked 18°.

In summation, conventional-type landing gear is superior for rough-field aircraft and working amphibians, but requires more pilot skill and technique, particularly during crosswind operation. In addition, a conventional gear in-

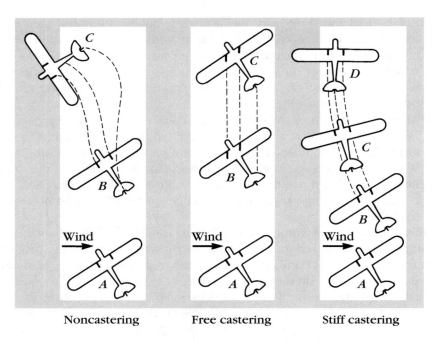

| Noncastering | Free castering | Stiff castering |

14-3 *The principle of self-castering main gear.*

stallation is lighter, cheaper, and requires less maintenance than a corresponding tricycle gear. So it is an ideal landing gear for a working airplane and professional pilots.

Tricycle gear

There certainly can be no doubt that tricycle landing gear has been a major factor in the expanding popularity and use of modern aircraft. Approach stability, longitudinal trim, and tricycle gear combine to make landing almost automatic, with ground handling no longer a major problem for the occasional pilot.

Figure 14-4 presents an average configuration for an acceptable tricycle landing gear design. Note that the tail clearance angle is based on the angle of attack for 90% of the maximum lift coefficient, since approach is always at some speed above stall. Ideally, the tail bumper or skeg should just clear the ground at this angle, or an angle 1 or 2° greater.

The main wheels should be positioned behind the most aft empty weight center of gravity so that the center of gravity always remains ahead of the wheel contact point; if the main wheels are too far forward the airplane will squat on its tail with the center of gravity behind the main gear. With these basic technical aspects of tricycle gear design in mind, let us review the practical benefits and problems.

Horizontal reference line   c.g.

Oleo

10° to 15°

$a = \Theta + 3°$

a

$D_1 \cong 0.67 D_2$

D_2

Ground line

Θ

$Y = 0.10 D_1$ (min)

X

Wheel base

Z

Θ = Tail angle for 0.90 C_L max, flaps down

$X = 0.09 Z$ to 0.15 Z

Tread = 0.75 Z to 1.25 Z

14-4 *Tricycle gear arrangement.*

Why does the tricycle configuration perform so well on the runway? As described in Fig. 14-5, the location of the main wheels behind and outboard of the center of gravity makes them tend to trail directly behind the center of gravity. This action is called inherent dynamic stability. We all know it is easier to pull a two-wheeled cart uphill than push it, particularly if the tow bar is pinned to the cart. And pushing a cart with a pinned bar is similar to the action of a conventional gear with free-swiveling tail wheel.

A   **B**   **C**

14-5 *Stability characteristics of tricycle landing gear. (a) Tracking down runway. (b) Recovery with a free-swiveling nose-wheel. Momentum acts to recover from a skid. (c) Windup possible with a steerable nose-wheel when holding rudder at touchdown to correct for sudden crosswind gust.*

Despite the rather obvious (when diagrammed) stability of the tricycle arrangement, a number of operational problems had to be resolved before tricycle gear became practical. Among those were the severe tendency for nose-wheel shimmy, the proper distribution of weight between the nose and main gear, the correct relationship between wing lift and main gear location, adequate tail power to rotate about the main gear at relatively low speeds, and reduction of installed landing gear weight.

All these factors do not have easy solutions; for example, it is impossible to obtain short-field takeoff performance from tricycle gear aircraft that will be superior to conventional gear designs of similar weight and wing area, a condition particularly apparent during rough-field operation, as previously noted. The Helio and Maule designs retain conventional gear for this reason, as do the Cessna 180 and 185, plus the fact that tricycle gear would have greater installed weight and cost.

Another problem inherent in the tricycle gear arrangement is shown in Fig. 14-5c, resulting from misuse of the steerable nosewheel. In fact, if the main gear or nosewheel doesn't collapse first, the condition shown will produce a ground loop as severe as any experienced with the most savage conventional gear. This problem used to occur in varying degrees with the two-control Ercoupe, which had interconnected rudder and ailerons plus a nosewheel steered by the control wheel for ground operation. Crosswind work had to be an approach banked upwind and carried well into the flare, with the nosewheel also turned upwind through the interconnected control system. If one didn't level out at just the proper moment, the plane either skidded downwind (recovery too early) or went smartly around upwind (recovery too late with the nosewheel still pointed upwind). As a safe alternative, the Ercoupe could be landed in a crab on the main gear with the nosewheel straightened before it contacted the ground.

Fortunately, the final approach procedure with three-control tricycle gear aircraft will normally be a bank into the wind with some top rudder; so even if the nosewheel and rudder steering are interconnected, runway contact will usually occur with the nosewheel aligned with the direction of motion, or at least that should be the intent. This also explains why modern nosewheels have spring bungee steering that allows the drag load to straighten any nosewheel offset angle present at runway contact.

Because most of today's flying is from paved runways, and because most of us prefer to be relaxed and humming away during final rather than "white knuckling" our prospects for ground-looping a severe crosswind landing, it isn't surprising that tricycle gear aircraft dominate general aviation use. But conventional gear designs will see continued service because this arrangement is superior for certain applications.

Whether tricycle or conventional gear is used, the use of wheel fairings will improve climb and cruise performance, as was shown in Fig. 6-8. In fact, the resulting climb performance halfway approaches that of fully retracted landing gear, but without as much maintenance, cost, and weight. Unfortunately, winter operation and wheel fairings are not particularly compatible in northern areas because of frozen slush and the possibility of locked wheels, so fairings are usually removed before the first snow.

Skis

As an alternative, skis can be installed.[1] Most ski planes are of conventional-gear type primarily due to the short-field performance offered. In addition, there is some skepticism about the use of skis on tricycle gear aircraft, although this arises largely through inadequate nosewheel design loads for snow operation rather than from any ground handling problems. The nosewheel strut and attachment structure are frequently much too weak to withstand frozen snow drift and ice ridge contact at high speed.

Tricycle gear aircraft intended for snow operation should be designed with all nosewheel strut and support structure load factors increased to twice the required landplane values. Although this will add somewhat to the airframe weight and cost, it is nothing compared to the inconvenience of being stranded in bush country at 20° below zero with a torn-out nosewheel and damaged prop.

Energy-absorbing systems

A brief look at the popular types of energy-absorbing systems will provide some practical background and useful design information. Also known as *shock struts* because they must withstand severe impacts, landing gear struts are designed to absorb dynamic loads by using one of the following energy absorption control methods:

- Shock cord
- Rubber discs
- Air and oil (also called *oleo strut*)
- Fluid alone
- Trailing beam with one of listed shock absorbers
- Springs, which can be simple leaf spring struts or compression spring/oil types

With the exception of the air and oil (or oleo) strut, all the methods dissipate very little energy by themselves, and what is absorbed is removed in the form of friction heat. Referring to Fig. 14-6 for each description:

- Shock cord is a simple and cheap way of controlling wheel deflection to absorb vertical velocity. But except for friction developed between elastic components, no impact energy is absorbed by shock cord. Of course, axle displacement permitted by the shock cord plus tire deflection reduce vertical impact velocity. This in turn reduces the landing gear load factor and airframe loading.[2]
- Rubber discs act like a shock cord, dissipating some energy by generating internal friction and possibly surface friction as well through contact with a cylinder wall. Again, the resulting wheel and tire deflections combine to reduce gear impact loads imposed on the airplane structure.

- Air and oil can be used to provide a desired rate of strut deflection by controlling flow through an orifice. The air chamber above the oil in Fig. 14-6c is pumped up to a pressure sufficient to support the static load on the strut when the airplane is at rest.

  Impact loads experienced during taxiing over rough ground or when landing compress the air and so increase strut deflection, forcing oil up through the orifice. This action dissipates (or absorbs) some energy while also controlling the rate of strut deflection, and so reduces landing loads by eliminating shock peaks.

  A low sink rate at wheel contact will require only a small deflection of the oleo strut to accept the landing load, which would be similar to the static loading; while a high sink rate at ground contact will cause large strut deflections at a rate controlled by the orifice design. To provide the low impact load factor desired, orifice characteristics must be developed and proven by drop tests.

  The metering pin shown might not be necessary to regulate orifice area vs. displacement for very light airplanes because a simple drilled hole can be adequate, but it is mandatory for heavier models due to the great variations in landing loads possible with larger aircraft.

  The oleo strut is an ideal method of controlling impact loads, as evidenced by its use on carrier-based and transport aircraft. Because an oleo strut requires more maintenance and is more expensive than simple spring leg struts, it is not as popular for small private aircraft, even though an oleo main gear should be lighter than a leaf spring steel strut.

- Although oil is considered to be a virtually incompressible fluid, Dowty developed a liquid spring strut many years ago based on the trailing beam design of Fig. 14-6d and e. As fluid pressure increases, the containing cylinder deflects, permitting the liquid strut to compress and the wheel to move up. However, the high pressures involved preclude use of this type of absorber on small aircraft because of the high cost and weight.

- The trailing beam strut might have shock cord, rubber discs, fluid, or an oleo strut for wheel displacement control as shown in Fig. 14-6e. The great advantage of this type of strut lies in its ability to smooth out rough terrain and equally rough landings. It also is a nice design for amphibian aircraft because it permits brake application on a wet ramp without causing main strut binding and resulting fore and aft chatter.

- The combination of oil, a compression spring instead of air, and an orifice used to be quite popular. This approach eliminated the need to monitor air pressure for correct operation, but the cylinder had to be fluid tight nonetheless. Unfortunately, coil compression springs tend to set with time and would frequently bottom with a bang if the oil level became too low, as it frequently did on the PT-19 that had this type of main strut. So progress has favored development of the leaf spring gear, whether it be of fiberglass, aluminum, or steel.

- Spring struts are very much with us today. Introduced by Steve Wittman and further developed by Cessna Aircraft, the leaf spring main gear strut of Fig. 1-3 is probably flying on more aircraft than any other type of landing gear. While leaf springs do not provide the energy control of an air/oil orifice system, they have the distinct advantages of extreme simplicity and virtually maintenance-free operation.
- Steel is the most effective material for leaf spring landing gear legs because of its strength characteristics. Since this type of strut is of interest to many owners and designers, an example of steel strut design is presented in the notes at the end of this chapter.[2] These calculations show how root section dimensions affect strut energy characteristics and so can vary the load factor used for landing gear and attachment structure design.

As a final comment regarding energy absorption, it is advisable to slant main gear struts slightly forward and down to align more closely with the combined vertical and drag load resultant, as shown in Fig. 14-7. Whether of conventional- or tricycle-gear type, you have probably noticed that most main struts incline forward. By doing this the designer has decreased bending loads on the strut and attachment fittings while also reducing the wearing loads on oleo struts.

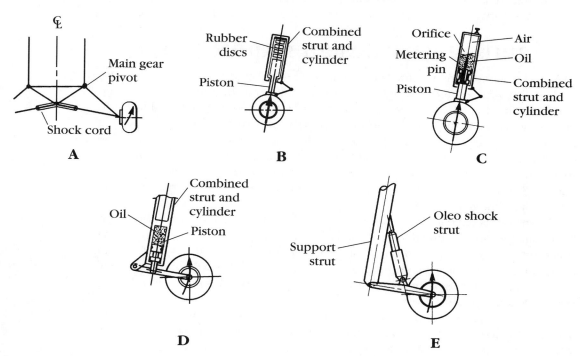

14-6 *Schematic drawings of various types of landing gear energy-absorbing systems. (a) Shock cord. (b) Rubber discs. (c) Oleo strut. (d) Liquid spring. (e) Trailing beam strut.*

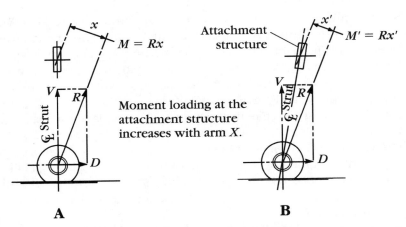

14-7 *Main gear load reactions. (a) Vertical strut. (b) Inclined strut.*

Although trailing beam shock absorbers are always axially loaded and so not subject to bending loads and friction, their attachment struts will be placed in bending by landing and brake loads; so even trailing beam gear legs should be inclined forward to reduce support strut and attachment bending loads.

Landing gear design requirements

FAA design conditions require each main wheel to be at least capable of supporting the airplane gross weight, thereby establishing the tire load rating and size.[3] That load must, of course, be multiplied by the 1.50 ultimate load factor to obtain the actual main gear design numbers. This means a 2400-lb gross weight airplane, whether of conventional- or tricycle-gear configuration, must be able to carry a load of 2400 × 1.50 = 3600 lbs through each wheel and landing gear strut into the airframe attachment structure. And sometimes even this is not enough strength for student landings; so trainers can have even larger load factors for landing gear design.

Nosewheels have to withstand about one-third the airplane weight both vertically and in an aft direction, with tailwheels subject to similar loadings. The horizontal loads anticipate obstructions such as bumps and furrows, but are not adequate for sustained rough-field operation or nose-ski installations as previously noted. Both nosewheel and tailwheel design factors should be doubled for continuous heavy-duty use.[3]

To prove just how rugged the final landing gear design is, as well as to substantiate the load conditions calculated for the main gear and nosewheel or tailwheel, FAA tests are again necessary. The specifications require a series of controlled drop tests, including one from 1.5 to 2 ft above the ground.[4] A loaded airplane can be used for these tests, or the individual main strut and tail or nosewheel units can be tested separately.

Either way, the strut being tested must be loaded to limit design conditions and dropped from increasing heights until the required distance has been reached—hopefully without breaking the wheel, strut, or any of the attaching bolts. The impact vs. time history of each drop is recorded and then measured to determine the actual load factor, as shown in Fig. 14-8. Most of the main gear drop tests will include impact onto an inclined wedge to simulate a fully braked or wheels-frozen landing condition, producing not only severe torsional loading of the main gear attachment structure but also some side effects in the data picture, as noted in Fig. 14-8.

14-8 *A typical drop test record.*

Despite the fact that all landing gear components plus the attachment structure have been thoroughly tested and are capable of taking the expected level of normal abuse, it is still advisable to check regularly the various wheels, struts, and structural connections to be sure nothing has worn excessively, worked loose, or started to crack out around bolts and bushings. The landing gear is probably subject to the most fatigue and impact loading of any part of the airplane, and should receive sufficient attention to ensure it doesn't literally let you down—embarrassing at best, as well as being very hard on the propeller, engine, wings, lower fuselage, and pocketbook.

Braking systems

Modern aircraft are fitted with hydraulic disc brakes because this type has less tendency to fade than drum brakes and will immediately wipe water, slush, and mud from the braking surface. A steel brake disc is fitted to the inside face of the cast magnesium or aluminum wheel, while the friction-producing brake lining material is fitted onto the facing of small pistons. Hydraulic pressure moves these pistons to clamp the brake disc in a caliperlike action whenever the brake master cylinder is operated in the cabin.

Wheel braking action can be applied (1) to both wheels at the same time by a single lever as used on Cherokees; (2) to either or both wheels by toe brakes actuated through rudder pedal deflection; and (3) independently or together by separate heel brakes. Toe brakes can be located only on the pilot side or at both front seat positions, while heel brakes are used in tandem arrangement on the Piper SuperCub; so we have a multiple choice of braking systems.

Fortunately, toe pedal action is rapidly becoming the standard method of brake application, with dual installations used to increase operational safety for instructional and demonstration aircraft. The single brake lever is fine when parking or to stop overrunning a taxiway or runway, but it offers no assistance when turning sharp corners or for independent wheel braking action frequently needed in high-wind conditions. Heel brakes are clumsy and difficult for the occasional pilot to operate, and they can be missed when heels start moving fast for urgently required braking action (to prevent a ground loop, for example).

In view of these deficiencies it is not surprising that as the number of private pilots increased, so did the demand for independent toe brakes. Piper has combined braking systems to offer both individual toe brakes and the single emergency/parking brake lever on the Cherokee series; so whether you are used to pulling a single lever or depressing toe pedals you can be sure of stopping in a hurry.

Parking brakes are now standard equipment on most modern aircraft and can be applied by (1) depressing and locking the toe brake master cylinders on the pilot's side; (2) depressing toe or heel pedals and locking the brake pressure in the lines by actuating a hydraulic locking valve; or by (3) pulling back and locking a single brake lever.

Either way, a control on or under the instrument panel will be pulled to engage the locking action, while the brakes will be released either by (1) releasing the locking lever; (2) depressing the toe or heel pedals; or (3) performing both actions at once. The problem with all parking being that hydraulic pressure is trapped in the brake lines to energize and lock the wheel brake pads onto each brake disc. If ambient temperature increases sharply after applying parking brakes, the hydraulic line pressure might also increase to a level capable of bursting the flex line or some other weak connection in the system.

To prevent such failures, hydraulic brake systems must be tested to 1.5 times the line pressure possible from the 200 lb maximum force a pilot can apply on the brake pedals.[5] But even this test does not assure freedom from ruptured lines at high temperatures, particularly in older aircraft. So if the day is going to be a hot one, chock your wheels and do not use the parking brake.

Retraction systems

While wheel fairings will increase climb rate and cruising speed, high-speed land planes must retract their landing gear to realize maximum performance. Of course, all amphibians must retract their gear prior to water operation.

Manual retraction is probably the simplest system and one requiring minimum maintenance. The actual retraction can be accomplished by direct hand lever action or by use of a hydraulic hand pump. The latter method has considerable merit for smaller aircraft if hydraulic pressure is stored in an accumulator until released to raise or lower the gear by actuating a selector valve.

Manual systems have another distinct advantage in that alternate retraction methods are not required by FAA regulations. If motor-driven hydraulic pump pressure or electrical retraction drives are used, it is necessary to supply a backup auxiliary hand pump or hand-crank method of lowering the gear in case the primary system fails. With hand power applied through a lever mechanism or pump handle we have only the basics, and this fact is recognized in FAA specifications.

Whenever the throttle is pulled back below normal base leg or approach power settings, land aircraft have lights and buzzers to warn that the gear is up. Some aircraft also have automatic gear extension systems that lower the gear whenever flight speeds get below-normal pattern levels. This system can be manually cut out for low-speed flight practice, but will be quite useless unless reactivated prior to entering final.

Amphibians are required to have similar warning systems even though the normal reaction to a flashing red light and horn signal would be to extend the gear during a water approach. As a result the number of amphibian landings on water with the gear down is sufficiently large to distort amphibian hull insurance rates, while the number of landings on land with the gear up is also distressingly large. The best answer for all aircraft operations is to establish a complete checklist for each phase of flight and then use it constantly, especially when flying an amphibian.

You have probably noticed that all aircraft usually retract the main gear inboard toward the wing root. This is done because outboard retraction places the gear weight further out in the wings, which in turn increases the airplane lateral mass moment, also known as lateral inertia. Increasing the lateral inertia results in either the installation of larger and more powerful ailerons to maintain a specified rate of roll, or acceptance of a reduced rolling rate for an established wing configuration. So for new designs and when converting an existing model from fixed to retractable gear, the main wheels are folded inboard whenever possible.

An excellent and complete review of many different landing gear types and designs, wheel and tire sizes, brake systems and installations, and similar useful landing gear design and service information is found in the notes listed at the end of this chapter.[6]

This chapter completes landplane design review. I hope the next chapter about seaplane design is of general interest to you because amphibian utility and safety have so much to offer all sport and business pilots. Following seaplane design, chapters covering the mysteries of aircraft protective finishes and FAA certification procedures have been included to complete your background from basic aircraft design through final approval.

Endnotes

[1](a) David B. Thurston, "North Country Skis," Aero Magazine, November/December 1975.

(b) AC 43.13-2A, pp, 33-41.

[2]Sample design calculations for an acceptable steel leaf spring main gear strut:

(a) FAA design conditions require each main wheel to carry a vertical load at least equal to the airplane gross weight per FAR 23.473(g) and FAR 23, Appendix C. This is equivalent to a 2 g impact load on the main wheel and strut. The following deflection formula will establish acceptable strut cross-section dimensions based upon calculations developed for the Teal amphibian and later substantiated by landing gear drop tests.

(b) Based on the accompanying sketch we have per Fig. 14-9:

Spring strut section A–A at fuselage side

6.00 × 6 wheel and tire

14-9 *Steel leaf spring strut nomenclature.*

Δ = strut deflection at the axle under 2 g impact load

$$= \frac{W\ell^3}{3EI} \times \text{Sec } \Theta', \text{ in inches}$$

where W = 2200 lbs for a 2200-lb gross airplane
   ℓ = 22.5 in. for this design
   E = the modulus of elasticity = 30×10^6 for steel (10×10^6 for aluminum alloy)
where I = moment of inertia of the strut root section = $b(h)^3/12$, in (in.)^4
   Θ' = 58° for this design

(c) Looking at the formula it is quite obvious that I holds the key to adequate deflection of the strut under impact load conditions. If I is too large, the deflection is too low and the impact load factor is high; in other words, the gear is too stiff. The following calculation shows the effect of deflection on the design load factor. When developing a strut design it is necessary to vary strut dimensions b and h until sufficient deflection is obtained to provide acceptable vertical load factors—and hopefully without a too soft or "springy" gear.

(d) If the strut selected is 3 in. wide at the base (b = 3) and 0.69 thick (h = 0.69), with $\Theta' = 58°$:

$$I = \frac{b(h)^3}{12} = \frac{3 \times (0.69)^3}{12} = 0.082 \text{ (per section A)}$$

Secant Θ' = secant 58° = 1.89

$$\Delta = \frac{2200 \times (22.5)^3}{3(30 \times 10^6)(0.082)} \times 1.89 = 6.41 \text{ in. strut deflection at the axle at 2 g impact}$$

(e) From 6.00 × 6 tire deflection charts at an inflation pressure of 20 psi, tire deflection = 3.14 in. at 2200 lb loading.

(f) Therefore, the total axle travel to descent velocity $V_d = 0$ is: 6.41 strut deflection + 3.14 tire deflection = 9.55 in. = 0.796 ft stopping distance = d.

(g) According to FAR 23.473(d), the initial descent velocity for landing gear design calculations cannot be less than $4.4(W/S)^{1/4}$, where W = gross weight and S = wing area. Assuming S = 157 sq ft, we have $V_{d_{min}}$ = 4.4(2200/157)$^{1/4}$ = 8.5 ft/sec. To find the landing gear vertical g design inertia factor: $(V_d)^2/2d$ = $(8.5)^2/(2)(0.796)$ = 45.4 ft/sec^2 and the vertical design inertia factor = 45.4/32.2 = 1.4 g due to stopping descent in 0.796 ft.

(h) The inertia load factor, η = 1.0 + 1.4 = 2.4 g (since the airplane and gear were subject to 1 g load in steady flight before the additional impact of 1.4 g was added at landing).

(i) Since FAR 23.473(e) permits a load reduction of 0.67 g due to wing lift assumed to act during landing, the landing gear design load factor becomes 2.40 – 0.67 = 1.73 g, indicating acceptable strut deflection.

(j) Because FAR 23.473(g) requires at least a 2 g limit design load factor, the landing gear limit design load factor used must be 2 g and not 1.73 g as indicated.

[3]FAR 23.471-23.509 and Appendix C.

[4]FAR 23.723-23.727.

[5]FAR 23.1435(b).

[6]Ladislao Pazmany, *Landing Gear Design for Light Aircraft*, Vol.1. Pazmany Aircraft Corp., San Diego, CA, 1986.

15

Seaplane design

A seaplane is an airplane capable of operating off water. Seaplanes can have twin floats or can be designed with a single main float plus small wing floats or similar lateral supports. The single-main-float seaplane is frequently referred to as a flying boat. And when any seaplane is equipped with wheels, it becomes an amphibian: an airplane capable of operating from land or water.

Various seaplane float, hull, and wheel configurations have been illustrated and discussed in chapter 4. So this review primarily covers seaplane design requirements, followed by some practical suggestions for successful and safe water operation that will be of interest to anyone flying off water.

Unfortunately, due to the few fixed-base operators catering to water aircraft, the pure seaplane is not as popular as it should be. For example, it is difficult for seaplanes to refuel when flying cross-country unless special plans are made in advance, or unless the seaplane is an amphibian capable of land operation.

Because amphibians combine land, water, and flight operational problems into one configuration, I will concentrate on that type. For seaplane design where only water operation is desired, simply remove the landing gear, and thereby increase useful load while also reducing purchase and maintenance costs.

The amphibian—
a safe and useful airplane

The primary reason for amphibian-type aircraft can be summed up in one word—utility. A very close secondary reason is safety. The amphibian is designed to operate from rough airports and available waterways, frequently without any prepared surface, docking, or handling facilities. As a result it is a most effective form of transportation into remote and undeveloped areas. Amphibian use will increase as emerging countries realize their economic potential. Fuel capacity for 6 to 10 hrs of cruise operation, coupled with landing versatility, permit the amphibian to fuel at home base, cruise outbound for 3 hrs or more to a remote camp or outpost, and return home without refueling. While in transit, every river, pond, or lake more than a half mile long is an emergency landing field, as are the usual open fields and intermediate airports. No other aircraft type encompasses such a broad spectrum of utility.

In addition, the rugged hull bottom resulting from water loads offers added protection in the event of a forced landing away from water or a suitable field, and this is a real safety item of benefit to every owner.

All of these features combine to offer a safe type of airplane for sports pilots as well. If they wish to reach a remote lakeside camp, amphibian pilots can fly from their home airport and land on the water at their dock. And when ski-equipped, a wintertime campsite landing may be made on the ice. Such capability provides year-round entry into inaccessible areas, and in all climates.

If the amphibian airplane is so versatile and capable, why isn't the type more popular? Because, as usual, there are several penalties accompanying the numerous advantages. The initial cost and weight of an amphibian compared with a landplane of similar capacity is greater because of the rugged hull and larger powerplant required. Secondly, operating costs are somewhat higher because of lower cruising speeds resulting from the hull shape, wing floats, engine location, and possibly the landing gear hanging out into the airstream. Useful load capability might be traded off against power vs. climb requirements, so an amphibian might not carry as much as is occasionally desired, especially on a hot day. And finally, some added maintenance results from the ability of water to find its way into almost every part of the airframe under the high pressures and velocities associated with water flight.

The amphibian is a compromise design combining water, land, and air operation into an acceptable vehicle. This is a challenging problem, with the unattainable solution being a configuration economically competitive with a landplane of similar power and gross weight. While water handling, power, and flotation design requirements preclude full realization of this goal, it can be nearly achieved through use of a surface-piercing hydrofoil positioned on a faired step or stepless hull, as discussed under hull design.

However, hydrofoil operation requires very demanding pilot technique and is also equally sensitive to variation in takeoff weight to such a degree that when lightly loaded the hydrofoil may unport before sufficient speed has been reached to permit aerodynamic control and stable flight about the hydrofoil (which acts as a pivot point below the airplane).

Seaplane nomenclature

The following nomenclature and related formulas apply to seaplane design.

    b   = maximum beam at the chine, ft
    c.b. = center of buoyancy

$C_{\Delta o}$ = gross load coefficient = Δ_o/wb^3, dimensionless
g = acceleration of gravity
H = step depth, ft
K = spray coefficient = Δ_o/wbL_f^2, dimensionless
L_a = length of afterbody in ft
L_f = length of forebody in ft
MAC = mean aerodynamic chord
n_{ws} = water landing impact load factor at step
$P_{H.T.}$ = horizontal tail power coefficient, dimensionless
$P_{V.T.}$ = vertical tail power coefficient, dimensionless
S_H = hydrofoil area in sq ft
$S_{H.T.}$ = horizontal tail area in sq ft
$S_{V.T.}$ = vertical tail area in sq ft
S_W = wing area in sq ft
V = velocity in knots
V_{so} = stalling speed in landing configuration, knots
w = specific wt. of water in lb/ft³: 62.4/ft³ freshwater, 64/ft³ seawater
W = gross weight, lb
y = distance from center line to float c.b., ft
α = sternpost angle, degrees (see Fig. 15-1)
β = angle of deadrise, keel to chine, degrees (see Fig. 15-1)
β_k = angle of deadrise at keel, degrees
Δ_o = gross displacement lb
Δ_{tf} = tip float displacement (freshwater), lb
τ = trim angle, degrees

Hull design

Although major components of any airplane are essential to its performance, hull design and resulting water handling characteristics are particularly crucial for an amphibian because it exists only because of its capability for water operation. In this regard, it should be noted that we have now reached a point in seaplane development permitting hull forms to be classified as conventional (having a transverse step) or advanced (either stepless or with a faired step). Each type is discussed separately in the following paragraphs.

Conventional hull design

The classic hull shape shown in Fig. 15-1 can be referred to as a planing tail hull because it only has one step with a forebody ahead of the step and an afterbody behind the step, each with V-shaped bottom surfaces for reducing water impact loads. Wing-mounted floats provide lateral stability when on the water. Conventional hull shapes and performance characteristics for small flying boats are

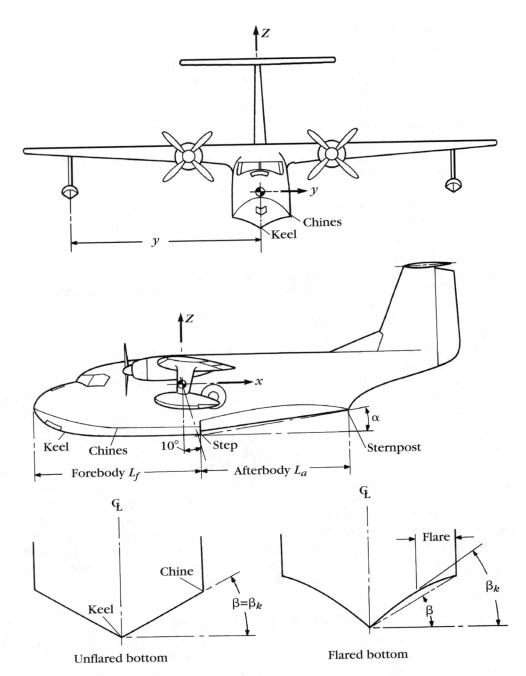

15-1 *Hull terminology.*

thoroughly evaluated and well presented in NACA TN 2503.[1] The trim advantages of slightly inclining the forebody keel 1 or 2° up and forward are apparent from TN 2503 data presented for Model 1057-04.

Forebody hull bottom deadrise angle β should not be less than 15° for the smallest hull and gross weight, increasing to 25° at the step for larger seaplanes in order to reduce design water load factors as shown in Fig. 15-2. Exceeding 20° deadrise, however, creates ground clearance problems for most configurations. As the bow is approached, deadrise can be increased to 40° by simply warping the bottom sheet on small hulls, while larger seaplanes benefit from reverse flare at the bow, although this requires stretched skins and is costly. In either case, increased deadrise at the bow reduces wave impact loading and increases sea-state capability for a given forebody length and beam.

Figure 15-2 also dramatically illustrates the important effect of V_{so} upon hull bottom loading. The design water load conditions are set forth in FAR Part 25.527(a) (1) and subsequent paragraphs for all commercial seaplanes, with Part 23 aircraft also covered in that section. The basic water loads formula is empirical and has been developed from operational experience:

$$\eta_{W_S} = \frac{0.012\, V_{so}^{\,2}}{\sqrt[3]{(\tan^2\beta)\, W}}$$

where N_{W_S} = water landing impact load factor at the step
V_{so} = stalling speed in knots (for all FAA water loads)
β = forebody deadrise angle, degrees
W = gross weight, lb

Because water loads are a dynamic impact function increasing with velocity squared, lowering landing speed by 30% reduces the bottom loading nearly 50% as shown by Fig. 15-2.

The values selected for deadrise angle and landing speed are crucial considerations for hull design because they affect the basic hull weight and cost as well as the sea-state capability. And a heavy hull is not only expensive, it also reduces useful load.

The basic relationship between forebody length, maximum beam, and beam loading is approximated in Fig. 15-3, which applies to floats and hulls. When calculating $C_{\Delta o}$ and K values for seaplanes less than 5000 lbs displacement and fitted with spray rails, the beam value b may be extended to cover the width over spray rails at the step. Figure 15-3 data shows a reduction in spray with increased forebody length; and because spray is a direct indication of hull drag, there is a natural tendency to extend the forebody. However, any excess vertical area ahead of the center of gravity further reduces the marginal directional stability inherent in all seaplane configurations, as discussed later under tail surface design. Therefore, it is advisable to keep the forebody length as short as passenger, cargo, and balance considerations permit with due allowance for a

15-2 *Step landing impact load factor variation with landing speed and deadrise angle for 4000 lb seaplane.*

reasonable hull planform at the bow. A forebody length-to-beam ratio of 2.75 to 3.00 should provide an acceptable compromise.

Based on these criteria, a 4000-lb seaplane hull 4.83-ft wide over the chines or spray rails at the step would have a 14.5-ft long forebody, a C_{Δ_o} value of 0.568, and a K value of 0.063, indicating satisfactory spray characteristics.

With the forebody length and beam established, the afterbody length and step depth may be determined by referring to Fig. 15-4. For a conventional hull, a deep step is preferable to a shallow one because the deeper offset in the hull bottom contour provides better step ventilation, which results in improved lift-off, less crucial trim limits, and reduced skipping and pitching tendencies. Additional ventilation may be realized by designing the beam at the step to be slightly narrower than that of the forebody maximum width ahead of the step. In other words, start bringing the hull in toward the sternpost ahead of the transverse step location. A step depth of 9% beam (at the chines) will be satisfactory for smaller seaplanes, reducing to 5% for hulls approaching 20,000-lbs displacement.

15-3 *Forebody length and beam vs. hull spray characteristics.*

To provide stable trim during planing, the hull step should be located on a line running approximately 10° aft from a keel perpendicular through the gross weight center of gravity (Fig. 15-1).

The afterbody sternpost and deadrise angles required for a stable hull are given in Figs. 15-5 and 15-6. This data is based on actual water handling characteristics of many seaplanes as well as towing test results obtained from model flying boat hulls. As a practical consideration, afterbody bottom warping should not be necessary for hulls less than 5000-lbs displacement; the afterbody dead-

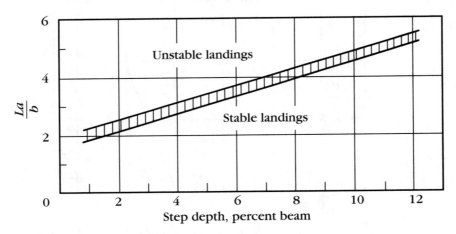

15-4 *Afterbody length and beam vs. step depth.*

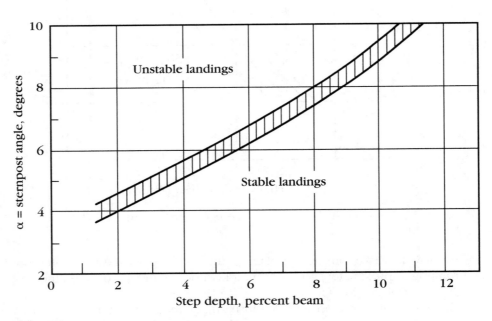

15-5 *Sternpost angle vs. step depth.*

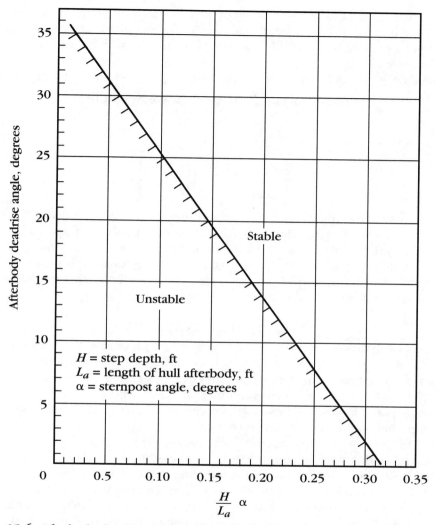

15-6 *Afterbody deadrise angle vs. length, step depth, and sternpost angle.*

rise angle may be the same as that of the forebody at the step. Also, the after-body angle must be adequate to keep the sternpost area clear of the water at the normal maximum hull trim angle τ of about 8°. A dragging afterbody or stern-post extends takeoff run at best, and may prevent takeoff of marginal perfor-mance seaplanes.

For planing tail hulls the afterbody should be slightly longer than the fore-body. Due to the tail arm required for flight stability, this relationship develops quite naturally when the tail surfaces are mounted at the sternpost as per mod-ern amphibian design practice.

In closing this discussion covering conventional hulls, some realistic values of sea-state capability are of practical use. There is a great tendency to expect too much of any hull, or to underestimate wave height, with equally unfortunate results in either case. Service reports indicate that the following wave height capabilities can be safely anticipated, subject to superimposed swell conditions:

Hull displacement in lbs	Wave height in ft
2,000	1
4,000	1.75–2
8,000	2.5
20,000	3.5
60,000	5
100,000	6

Because waves have been known to break over the flight deck of aircraft carriers, the use of amphibians and seaplanes for open ocean operation is restricted to large displacement hulls and acceptable surface conditions. However, the sea-state capability of any flying boat hull may be considerably increased by the addition of a retractable hydrofoil extending from the keel, a new development briefly reviewed in the following section.

Next generation hulls

A series of hydro-ski and hydrofoil development programs was sponsored by the Naval Air Systems Command and conducted by the author through 1970. This work resulted in the successful flight evaluation of a single, surface-piercing hydrofoil mounted on a conventional hull as an effective means of permitting a seaplane to operate in sea-state conditions beyond its normal capability. A comprehensive review of this program as well as hydrofoil seaplane detail design data are presented in the notes at the end of this chapter.[2]

Carrying this development further, it is possible to combine the hydrofoil with a faired step hull as a means of reducing cruise drag. Continued improvement through rounded or faired chines in combination with a faired step or a stepless hull design would reduce flight drag to a probable minimum for an amphibian. Such an advance must be thoroughly investigated by towing basin tests to prove the ability of the hull to unstick at hydrofoil unporting speeds and then to trim properly when planing on the foil.

The actual hydrofoil area required to support a 4000-lb displacement amphibian at 30 knots may be determined through the following typical calculation:

$W = \Delta_o = 4000$ lbs
$S_W = 260$ sq ft (slotted flaps)
Hull trim angle prior to unporting hydrofoil = 7.5°
Wing $C_L = 1.9$ (wing angle of attack is sum of wing incidence angle plus hull trim angle)
Hydrofoil unporting speed = 30 knots

Note that the unporting speed must be high enough to provide effective aerodynamic control. If the hull becomes foilborne at too low a speed, the seaplane becomes uncontrollable due to lack of control surface effectiveness.

(i) Wing lift at 30 knots = $0.0034\ C_L S_W (V_{knots})^2$

$$= 0.0034 \times 1.9 \times 260 \times (30)^2$$

$$L_W = 1510 \text{ lbs}$$

(ii) Hydrofoil lift required = Δ_o − wing lift

$$L_H = 4000 - 1510 - 2490 \text{ lbs}$$

(iii) S_H = hydrofoil area in sq ft = $\dfrac{L_H}{1.424\ C_L \rho_w (V_{knots})^2}$

where L_H = hydrofoil lift, lb

C_{LH} = hydrofoil lift coef. = 0.27

(Note: this 0.27 value is about maximum for a supercavitating hydrofoil. Foil angle of attack for this lift coefficient is 7.5°, the same as hull trim angle. This foil is therefore set at 0° incidence to keel.)

For this example:

$$S_H = \frac{2490}{1.424 \times 0.27 \times 1.94 \times (30)^2} = 3.71 \text{ sq ft of hydrofoil planform area}$$

ρ_w = water density = 1.94 for freshwater

= 1.99 for saltwater

V = hydrofoil unporting speed, knots

The density of freshwater is generally used with this equation to assure adequate foil area for freshwater operation. The relatively small area required to provide 2490 lbs of support at 30 knots is due to the fact that freshwater is about 815 times the density of standard-day sea-level atmosphere, which in turn explains why the hull bottom and support structure must be very rugged.

Detailed design data for a suitable hydrofoil section, support strut, and position along the hull bottom are covered in the notes at the end of the chapter and are not discussed further in this book.[2] However, it is interesting to note that a basic hull capable of handling 2-ft waves would be expected to operate safely in 4-ft seas when supported upon a hydrofoil extended 2 ft below the keel. The first successful working example of this concept, the HRV-1 hydrofoil amphibian, is shown planing on a single supercavitating foil in Fig. 15-7a. The flight photo of Fig. 15-7b was taken during climb-out immediately after takeoff, offering a comparative view of required hydrofoil size and shape.

A

B

Howard Levy

15-7 *The HRV-1 test amphibian (a) planing on a supercavitating hydrofoil, and (b) taking off with hydrofoil extended.*

Wing design

The wing and any related lift-increasing devices such as flaps or drooping ailerons should be designed to provide the minimum landing speed that is economically practical. The ability to approach safely at 60 knots and touch down at 45 knots opens many small bodies of water and small fields for possible landing sites. Further, at low stalling speed the water takeoff time and run will be considerably reduced compared with a design that must accelerate to a higher breakout speed. And reduced water contact time and speed mean safer operation in rough sea conditions.

After takeoff, climb becomes crucial. Wing area and horsepower provide climb, but high aspect ratio wings have comparatively low induced drag at low (climb) speed and so offer increased ROC for a given wing area and power loading, as was covered at length in chapter 6. In view of this, wing aspect ratios of 8 to 10 should be considered; the slightly greater weight of a high aspect ratio wing structure is more than compensated for by increased performance throughout the flight envelope.

The two parameters of wing area and aspect ratio establish the wing planform geometry, which should include a slotted or similar type of lift flap to provide an acceptable lift coefficient. High drag devices such as split or plain flaps should be avoided because amphibians have more than enough drag for steep approaches. Flaps generally improve water takeoff performance and should be designed to remain as clear of solid water as possible. Water impingement increases takeoff drag, extending takeoff run and time. Repeated spray contact can also damage light structures, so flaps must be designed taking into account water operational requirements.

The wing airfoil should be selected to provide minimum aerodynamic moment and drag. Hull shape and thrust line location alone usually require considerable power "off" longitudinal trim capability to meet FAA low-speed stability criteria, without having added amounts of trim required by the airfoil section. This is particularly true at approach speeds in the landing configuration with flaps and gear down (power off). Since selection of suitable airfoils has been covered in chapter 6, this information will not be repeated here.

A design relationship peculiar to the amphibian occurs in the selection of wing angle of incidence relative to the hull keel. As mentioned under hull design, the keel should be inclined up and forward 1 or 2° at the step to broaden trim limits. When planing on the step, the hull will trim with the keel inclined about 8° to the water surface. Since maximum lift should occur with the hull at this attitude, wing incidence will be established accordingly, depending on airfoil characteristics and takeoff flap deflection. Note that land-operating attitude requirements have to be made compatible with water-operating criteria; this occasionally becomes a challenging compromise because the afterbody sternpost must remain just clear of the water surface but also must miss the airport runway during full stall landings. Because the wing incidence angle established for water lift-off usually is too high for normal cruising flight trim attitude, amphibians tend to fly nose low during cruise. This improves visibility but requires some familiarization for level flight trim. The use of flaps tends to shift maximum lift to a lower angle of attack which, fortunately, reduces wing incidence and so improves hull trim during flaps retracted cruising flight.

Ailerons should be designed for minimum adverse yaw because flying boat directional stability tends to be marginal at best, as discussed in the following section. The ailerons must also be capable of lifting a low wing at a speed of about 20 knots in order to permit crosswind takeoff from water. Larger than normal ailerons are necessary to satisfy this requirement, with the usual result that they feel stiff in cruise flight. This condition may be improved by using a relatively thin (9 to 10%) airfoil section ahead of the aileron area. As with flaps, water impingement must be considered as an aileron design condition.

If specifications permit, wing area should be based on 0-9 $C_{L_{max}}$ to permit future growth with acceptable takeoff speeds. For comparative assistance during preliminary design, average values of wing loading variation with gross weight have been plotted for single and multiengine amphibians in Fig. 15-8.

15-8 *Wing loading vs. gross weight, based upon 46 single and multi-engine seaplanes.*

Tail surface design

The T-tail arrangement with the stabilizer and elevator positioned at the top of the fin provides excellent spray clearance for the horizontal surfaces. With single-engine amphibians of this design, the horizontal tail may be located in the propeller slipstream to obtain minimum trim change with different power settings. In this position the horizontal tail also remains clear of flow changes due to flaps, although buffet warning may be quite light at the stall.

With the stabilizer acting as an end plate atop the fin, the vertical surface is more effective for a given area. Although this improves directional stability, fin design loads and the horizontal tail unsymmetrical loading will be slightly greater due to horizontal and vertical surface interaction in lateral gust and maneuvering conditions, as discussed in chapter 7. However, any increased directional effectiveness is very important because hull-type amphibian configurations tend to be marginally stable directionally due to the bow area ahead of the center of gravity. Not only does the hull profile move the center of lateral area close to the center of gravity, but as shown by Fig. 15-9 the bow shape will produce a lift component in yawed flight that tends to hold the amphibian in a displaced, yawed attitude during a skid, sideslip, or lateral gust (turbulent) condition. This

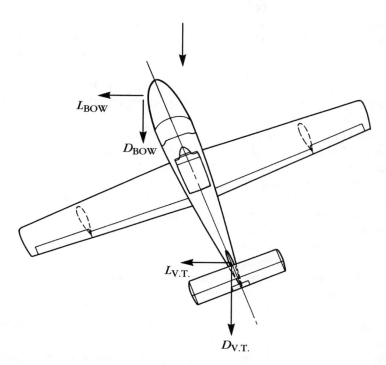

15-9 *Bow lift in yawed flight condition.*

inherent hull design problem may not be generally recognized, but it does exist and must be considered during preliminary design.

For seaplane design the horizontal tail must be able to trim large moment changes caused by a relatively high thrust line acting about the center of gravity, hull drag, step drag and location below the center of gravity, and flap application—all of which cause tail-load variations with speed and power changes. For larger amphibians an adjustable stabilizer is recommended, although the drive and pivot attachment must be ruggedly designed to withstand rough-sea operation, stall buffet, and possibly propeller slipstream.

Nose-up longitudinal trim with power "off" and flaps down may be crucial at the forward center of gravity gross weight condition. If so, elevator tab effectiveness will be increased by static balancing the elevators or by adding a balance spring into the elevator control system; either method will overcome the natural tendency of an unbalanced elevator to trim trailing edge down in flight. Design criteria for adequate and effective tail area is found in chapter 7.

Landing gear design

The type of landing gear, conventional or tricycle, and main wheel location are important basic considerations for amphibian landing gear design. The final configuration will require many compromises because the landing gear must:

- Have structural support adequate to carry land and water landing loads, although the heavily loaded hull or wing structure must be cut away to accept the retracted gear
- Provide land landing clearance for the afterbody hull sternpost, yet the main gear support strut length must not cause major structural design problems
- Be capable of land operation without nosing over when braked at full throttle (conventional gear) or tail squatting when empty (tricycle gear)
- Be stowed or raised clear for water operation, but be immediately available for beaching or ramp approach

Amphibian pilots operating in bush country and remote areas prefer conventional (main wheels forward plus tailwheel) landing gear because it is easier to handle when leaving or entering the water and provides better rough-field operation than tricycle gear. These advantages have been covered in chapters 6 and 14 and illustrated in Fig. 6-1.

Although conventional gear is preferred for utility operation, high-speed amphibians intended primarily for the executive and sports markets should be equipped with tricycle gear for easier handling on land. This type of amphibian experiences less frequent water operation than utility models and should be designed more for land service.

If at all possible, develop the basic amphibian structural layout to accept both conventional and tricycle landing gear installations. Depending on the intended design use, it should be possible to concentrate initially on one landing gear system for early production, and then develop the other type as the market expands.

The problem of landing on the water with the gear down and on land with the gear up cannot be overstated. Despite attempts to tie checklists into warning light systems through color coordination, the percentage of amphibian landings with the gear in the wrong position remains high. Using an amber light and amber checklist background for water landings (gear-up) and a green light and green checklist for land operation (gear-down) should help, but only if cockpit procedures include reading the checklist prior to landing.

Structural damage from gear-up landings on land can be reduced if a steel skeg or skid wear plate is fitted to the keel at the step. The step area is designed to support such loads due to water pressure requirements.

Water landings with the gear down generally cause the most damage and can be fatal unless the hull is extremely rugged. Repairs can be reduced by designing the main strut attachment to withstand the torsional impact loading of a wheels-down water landing, but at a weight penalty that seems necessary for a safe and acceptable amphibian design.

Powerplant design

To provide required service, a utility amphibian must have sufficient thrust to assure rapid acceleration and climb-out at design gross weight on a 90°F day.

Rapid acceleration reduces water takeoff time and distance necessary to reach breakout speed—usually about 5 knots above stall for a reasonable margin of control. As previously mentioned, minimum water contact time is a safety feature that reduces wave impact exposure in rough-sea conditions as well as the possibility of striking floating or partially submerged debris. A gross weight climb rate of at least 800 ft/min under standard conditions is desirable to clear wooded shores, to overcome downdrafts, and to clear hills usually associated with rivers or lakes. The design objective is to reduce exposure to both wave impacts on the water and mountainsides when airborne.

To satisfy these criteria the design static thrust available should exceed a value of one-third gross weight. At an average static thrust of 4 lb/hp, a minimum of 330 hp is required for a 4000-lb gross weight amphibian, representing a power loading of 12 lb/hp. Typical values of power loading vs. gross weight are plotted in Fig. 15-10 for actual single and multiengine flying boats and amphibians.

It is obvious that the propeller or propellers should be located clear of all heavy spray and solid water thrown out by the hull. However, this requirement has been too frequently ignored in the past with the result that excessive propeller maintenance and replacement have created customer dissatisfaction with water operation in general. Protecting the propellers requires a raised powerplant installation whether the configuration is a single or multiengine design, while a high thrust line means fairly large trim changes with power. In view of this, three-bladed propellers should be used with larger engines to keep the

15-10 *Power loading vs. gross weight, based on 46 seaplanes.*

thrust line as low as possible because they deliver power from a smaller diameter than two-bladed types. The three-bladed propeller is also quieter because of reduced diameter with lower tip speed at full power rpm.

For larger amphibians a reversing pitch propeller should be provided for holding position in a seaway, backing down from a dock or beach, and braking dock approaches in adverse wind and current conditions. This feature is not mandatory for small amphibians because a paddle usually provides adequate docking assistance.

To keep the single-engine amphibian propeller clear of spray requires either a pylon-mounted tractor or pusher nacelle located above the hull near the fore and aft center of gravity location; a tractor installation mounted ahead of the windshield over the bow; or a spray-rail-protected pusher assembly faired into the aft cabin section, which can experience considerable propeller erosion. For single-engine amphibians, the tractor installation delivers more thrust in flight than the pusher design while also providing some hull lift if the wing center section or similar hull top contour is exposed to slipstream flow, as seen in the Teal amphibian of Fig. 4-13.

As an item of design interest, the one-engine-out directional control problems usually experienced with twin-engine designs can be virtually eliminated by using a tandem tractor-pusher installation. This package could be located above the hull centerline in a pylon-mounted "twin pack" nacelle.

If sound is to be reduced to the levels necessary to satisfy current FAA environmental (noise) specifications, all new designs whether of single or multi-engine configuration will require the lowest possible propeller tip speed combined with the use of effective mufflers.

Transverse stability requirements

Flying-boat designers have used hull side sponsons, immersed wing tips, planing skis, and similar approaches to obtain transverse stability on the water. While such appendages are effective when underway, they must also provide static stability—and this requires adequate physical displacement.

For smaller amphibians, a wing float located near the tip is most satisfactory. To accommodate the weight of a person walking along the wing, which occasionally might be the only way to disembark, the tip float buoyancy should be sufficient to support a 200-lb load at the wing tip. This condition is usually crucial for small flying boats, although the effect of wind side loads of 35 knots should be investigated for the design at rest in the water.

Although there are many complex formulas available to compute support float volume, for seaplanes less than 8000 lb gross weight a suitable displacement value for each tip float will be obtained from the relationship

$$\Delta_{tf} = \frac{.75\,W}{y} \text{ to } \frac{1.25\,W}{y}$$

where Δ_{tf} = freshwater displacement of each tip float, lb
   W = gross weight of amphibian in lbs
   y = distance from centerline to float center of buoyancy in ft

The tip float bottom surface should provide dynamic lift when running on the step to permit banking into the direction of turn without submerging the inside wing panel. By reducing the tendency to skid out of a turn, banking permits water turns of smaller radius and at higher speeds than would be possible with tip floats that could submerge underway.

Wing floats should be designed with a structural fuse attachment that will shear off when striking a log or similar object in the water, protecting the supporting wing structure from basic damage.

For high-performance amphibians, wing floats can be retracted in flight to act as end plates located outboard of the ailerons. But they must be designed to extend and retract at the same time to avoid directional stability problems. Further, note that this method of float storage reduces available wing area as floats are lowered; just when maximum area is desired to reduce landing speed.

Operational requirements

Although entire books have been written about seaplane handling and operational requirements, designers are mainly concerned with the safety and utility aspects of their particular configuration. A few design considerations not previously covered are:

1 Because an owner purchases an amphibian primarily because of its ability to operate from water, any design compromise should be resolved in favor of improved water handling characteristics, assuming pilot safety is not involved. This is a basic consideration for any amphibian design.

2 The importance of designing for the minimum possible landing speed cannot be overstated. As shown dramatically by Fig. 15-2, reducing the landing contact speed from 70 knots to 50 knots cuts the water impact load factor virtually in half. While this is not surprising since impact magnitude is a dynamic function of velocity squared, it is extremely significant from the safety standpoint. And not only during landing. Any reduction in gross weight stalling speed is accompanied by lower takeoff speed, which in turn reduces takeoff run and time. Shorter takeoff run permits safe access to smaller bodies of water, while reducing water takeoff time minimizes contact exposure to rough seas and possible floating objects. Coupled with normally better hull characteristics at lower speeds, all factors associated with reduced landing speed act to increase water operational safety.

3 The amphibian is a utility airplane and hopefully a workhorse for serving remote areas. As a result, the ability to load and unload relatively dense cargoes such as packaged foods, chemicals, fuel drums, and sick people must be investigated and resolved during preliminary design.

4 Fuel quantity is another important consideration. Flight into remote areas usually means no refueling until return to home base. Depending on the size of the amphibian as well as the operational specifications, fuel capacity for a minimum flight time of 5 hrs should be provided, with optional capacity available for 8- to 10-hr flight at maximum-range

cruise power. In addition to extending the radius of action to useful limits, such duration is a source of comfort during IFR conditions and permits extended flights between established airports where fueling is possible. Fuel should be stored in wing tanks outboard of the hull; considerable structural relief will be realized at gross weight with full fuel in this location while passengers will be clear of fuel if structural damage occurs.

5 Docking, particularly in a seaway, strong current, or on a lee shore requires at least minimum protected facilities. A suitable dock for remote areas can be obtained by constructing a horseshoelike structure moored from the center of the closed end. This structure will stream with the open end downwind or down current, permitting the flying boat to taxi into a containing dock with off-loading ramps positioned either side of the hull as shown by Fig. 15-11. Such a mooring is almost mandatory for rocky and steep shores or beachless areas.

15-11 *Horseshoe dock design.*

It is unwise to leave any type of seaplane afloat at anchor by itself for overnight or longer; the hull plating could have been pierced by floating debris or a structural joint might have opened during the last landing. If the amphibian cannot be taxied ashore or beached for extended periods, it should be docked in such a manner that it cannot sink. While automatic electric bailers might provide some mental relief for this condition, their starting float can stick or the aircraft battery can fail. At best, integral bailers are a secondary safety feature.

6 Mooring cleats located near the bow and tail are essential for water handling, mooring dockside, and anchoring. Each cleat and its supporting structure should be designed to withstand shear and tension loads separately applied (not combined), but equal to the amphibian gross weight. Although mooring line loads would never be expected to reach this value, an anchored amphibian will hook, snatch, dance, and generally work on the mooring line just like any boat, developing relatively large inertia load factors at the attachment cleat. In addition, the normal wing-tie-down provisions should be included for securing on land.

7 Last, but as basic as any other design consideration, is the need to provide corrosion protection throughout the airframe. The amphibian must be able to operate off salt water with such minimum maintenance as washing with fresh water after each flight and regularly filling rotating joints with grease to keep out all water. This means alodining and priming all aluminum alloy parts; cadmium-plating steel pieces; sealing exposed skin joints, beam caps and doublers, and major structural attachments; using aluminum alloy wheels and brakes; stainless steel control cables; and isolating dissimilar materials.

Operational procedures

For maximum safety, the following operational procedures should be followed.

1 Most small seaplanes receive FAA approval for operation in 6- to 12-in wave conditions (as measured from trough to crest). Although some aircraft can be safely flown off 24-in high waves, operation of a small seaplane in seas of this size is not recommended for anyone except the very experienced water pilot.

   In addition to the usual surface waves, the mouth of a sheltered bay opening onto the ocean usually contains running swells that can combine with local wave conditions to provide extreme pitchup during a takeoff run into the advancing swells. Such surface conditions can result in becoming airborne prior to attaining flying speed and should be countered by running parallel to the wave front.

   Note that the wake of a powerboat can produce superimposed swells of very short wavelength that, at the least, result in a very uncomfortable takeoff run. Depending on the size of the powerboat, such swells can also pitch a small seaplane into the air before reaching minimum flight speed.

   When experiencing steeper seas than anticipated during takeoff, abort the run by chopping throttle and turning crosswind (to the left or right, whichever direction is toward shore and calmer water). Because of the stabilizing action of the wing floats on midwing hull-type seaplanes, they will not capsize as would a pontoon seaplane caught in a high crosswind.

2 When judging surface conditions from approximately 500 ft above the water area intended for landing, it is suggested that the observer multiply the wave height estimate by a factor of 2. Water has a peculiar habit of growing in wave height the closer an airplane approaches the surface during final. As a general rule, large unprotected and open lakes are not suitable for water operation far from shore when surface winds exceed 10 mph. Although prior experience has shown that seaplanes can be full stall landed on water in winds up to 30 mph (the surface wind speed reduces the water contact speed by a like amount), during such wind conditions large lakes can develop very steep seas (short wavelength from crest to crest) that make takeoff quite dangerous, if not impossible.

By contrast, when operating from the usual small lakes up to 1 × 2 mi in size, there is not sufficient distance (fetch) for large seas to develop; with these conditions, water flight can be safely conducted in winds up to 25 to 30 mph. Even so, the takeoff run can be uncomfortable.

3 Always plan your approach to land on the windward (direction the wind comes from) side or end of the lake or bay. Through this procedure, you will contact seas that have minimum fetch with correspondingly less wave height, as well as being near a protected shore if the sea state is considered too great for takeoff. Possible landing areas are shown on Fig. 15-12.

4 If caught in heavy seas after landing, do not attempt to take off. Taxi toward the windward shore (the shore upwind of the airplane) at about 10 mph with hatches closed. If calmer seas are reached, takeoff can be made. It might also be possible to taxi into the lee of a point or behind an island or breakwater providing calmer sea conditions that permit takeoff (Fig. 15-12).

5 When it is not possible to find acceptable water conditions, taxi up onto a beach or moor the airplane on the downwind side of a dock or float located on the windward side of the lake or bay. Remember, when on the water a seaplane is a boat and should be handled as one. So when mooring be sure to attach bow and stern lines, as well as spring lines fore and aft to prevent airplane motion relative to the dock, as shown in Fig. 15-13a. Leave sufficient slack to permit the airplane to float at least 1 ft away from the dock (the wind holds it away from the downwind side of a dock or float). If suitable line or docking cleats are not available, moor the plane with a single yoke line around both bow cleats and running to the downwind side of the dock; the surface wind will stream the seaplane away from the dock and the fin area will keep it headed into the wind weathervane fashion.

15-12 *Water areas for safe operation in heavy wind conditions.*

When no docking or beaching is possible, taxi close to the windward (upwind) shore, fasten a mooring line around the bow cleats to form a yoke leading to a single anchor line, and throw over the anchor. Numerous small boats will shortly appear to offer rescue, and can cause more problems than the wind conditions ever could.

6 As a final note, water flights should be conducted with all aboard wearing inflatable life vests, and with an anchor line, anchor, and paddle firmly secured in the cabin. For a more detailed and comprehensive review of water flying, Frank Kurt's *Water Flying* is recommended.[3]

15-13 *Possible docking and mooring conditions. (a) Taxi to dock and handle to positions shown. (b) Single-point mooring for short time.*

Endnotes

[1]W. C. Hugli, Jr., and W. C. Axt, "Hydrodynamic Investigation of a Series of Hull Models Suitable for Small Flying Boats and Amphibians." *NACA Technical Note 2503*, November 1951. (Copies available from National Technical Information Service, Springfield, Virginia 22151.)

[2]D. B. Thurston and N. J. Vagianos, "Hydrofoil Seaplane Design." Unclassified Final Report No. 6912 for Naval Air Systems Command under contract N00019-69-0475, May 1970.

[3]Franklin T. Kurt, *Water Flying*. The Macmillan Company, New York, 1974.

16

Materials and finishes

Many years ago, when the application of pure aluminum coatings to both surfaces of aluminum alloy sheet was first introduced as the "alclad" process, a shiny airplane was the height of fashion. When weather did not permit flights around and between airports, time could always be spent leisurely polishing the clad wings, fuselage, tail, and cowling while accepting compliments and discussing the latest airport news.

As spare time became ever more scarce and labor rates soared accordingly, the novelty of polished surfaces rapidly faded in favor of painted aircraft. This change in style was also fostered by the need to improve the visibility of increasing numbers of aircraft operating in clear sky or haze conditions because an alclad finish tends to blend into its surroundings. It also became evident with time that (1) cladding would not stop corrosion over large scratched or abraded areas and (2) continued polishing would eventually wear away the protective surface, which, in either case, had to be painted or replaced.

So a whole field of protective processes and coatings evolved for aircraft use, with development and application rapidly accelerated by World War II camouflage requirements. Because shiny airplanes at altitude reflected sunlight like flying mirrors, they had to be given a lasting, nonreflective finish, with weather-resistant, fuelproof paints becoming the solution to this problem.

Although the basic corrosion protection process differs for steel and aluminum alloys, the same primers and finish paints can be applied to either material. Starting from the inside and working our way out to the surface, let us first briefly look at the different materials used for airframe construction and then see how they must be treated to provide the long life expected from modern aircraft.

Aircraft materials

Table 16-1 has been prepared to indicate the different aluminum and steel alloy strengths available and their accompanying Rockwell numbers. The Rockwell number is a calibrated reading obtained when a metal surface is penetrated by a small steel ball or shaped diamond point loaded at a known value. The amount of penetration is an indication of material hardness that can be instantly read from a calibrated dial.

Table 16-1 Materials and their hardness—aluminum and steel alloys commonly used in aircraft construction

Material	Ultimate tensile stress, psi	Material hardness Rockwell B 100 kg	Rockwell C 150 kg
		Aluminum alloys	
5052-H34	34,000	B15/B22	
5052-H38	39,000	B28/B38	
6061-T4	30,000	B3/B15	
6061-T6	42,000	B38/B45	
2024-T3	Bare 64,000	B72/B76	
	Alclad 59,000	B67/B72	
7075-T6	77,000	B82/B87	
		Steel alloys	
1015-1025 mild steel	55,000	B64/B68	
4130A (annealed)	70-85,000	B77/B87	
4130N (normalized)	90-95,000	B90/B94	
4130-heat treated	125,000		C26/C30
4130-heat treated	150,000 max.		C33/C37
4340-heat treated	180,000		C39/C42

For example, the B-100 kilogram listing indicates a $\frac{1}{16}$ in. diameter size B hardened steel ball is pressed down upon a metal surface by a 100-kg force. If this ball penetrates the surface sufficiently far to produce the low reading of B20 on the indicator dial, we know from Table 16-1 that the material being checked is 5052-H34 aluminum alloy. In a similar manner, if the penetration is only slight, say to a B85 reading, we know 7075-T6 aluminum alloy is being tested. Steel requires a sharper penetrator for the higher-strength alloys, so a Type C diamond point is used under a load of 150 kg (the right-hand column of Table 16-1).

Through this nondestructive inspection procedure it is possible to identify various materials and check completed items to ensure they are of the correct material, and to do so without harming the part. FAA inspection procedures require that every piece or lot of parts be checked in this manner for proper material strength, and also that certified records be maintained to document such inspection by airplane number. Through this process it is possible to determine which run of parts was used on each airplane if something should go wrong at a later date. Naturally, all this adds to aircraft cost, but also to flight safety. A more detailed review of nondestructing testing is in the notes at the end of this chapter.[1]

Aluminum alloy is the principal material used for aircraft construction for reasons indicated by Tables 16-1 and 16-2. Table 16-1 shows that 1015-1025 mild steel is not as strong as 2024-T3 bare or clad aluminum alloy sheet, although it is 3 times as heavy as shown by the weights given in Table 16-2. As a result, mild steel such as 1020 hot- or cold-rolled stock should not be used for aircraft parts because of its poor strength-to-weight ratio. By comparison, alclad 2024-T3 sheet offers about two-thirds the strength of normalized 4130 alloy steel at only one-third the weight.

Aluminum alloys are used in many forms. The most common form being sheet made into skins, ribs, and frames, with extrusions used in quantity for beam caps, reinforcement angles, and small fittings tying parts together. Drawn tubing, rolled bar, special forgings, and castings are also fabricated from various alloys suitable for the different forming operations as listed in Table 16-3. For the highest strength-to-weight ratio aluminum alloy structures, bare sheet should be used because cladding material is of low strength, reducing allowable stress levels for clad sheet.

Extrusions cannot be clad because these shapes are formed by forcing aluminum that has been heated to a plastic state through a die of the desired cross section—much like making spaghetti. After being cut to length, each extruded piece is straightened by pulling on the ends the way string can be pulled tight (and straight).

Aluminum alloy tubing can be drawn or extruded. Drawn tubing is accurately formed by being worked to size through a series of forming dies. This type of tubing is seamless and stronger than extruded tubing, with the latter used primarily for noncritical secondary structures or nonaircraft commercial applications.

Heat-treatable steels are used in aircraft construction when space and the superior fatigue qualities of steel are crucial considerations. An example of this is wing attachment fittings that must be located as far apart between the top and bottom beam caps as possible, yet not protrude beyond the wing section contour.

Normalized 4130 steel is the alloy usually used for light aircraft steel parts and fittings, having a useful ultimate strength of around 90,000 psi as shown in Table 16-1. Fortunately, the strength of 4130 steel can be increased by mild heat treating to 125,000 psi; at which level it is twice as strong as 2024-T3 alclad, but still weighs 3 times as much. But any further increase in 4130 strength requires holding fixtures during heat treating, results in parts and assemblies difficult to drill or rework, and might produce brittle structures.

The highest heat treat that should ever be considered for 4130 steel sheet, tube, or bar parts is 150,000 psi. At this value steel will be almost 3 times the strength of clad 2024-T3 at 3 times the weight. This trade-off is accompanied

Table 16-2 Comparative properties
of aircraft materials (see Fig. 16-3 for material specifications)

Material	2024-T3 Alclad sht.	2024-T4 extrusion	Normalized 4130 sheet		4130 H.T. to 125,000 psi
Thickness →	.010–.249	.050–.249	≦.187	>.187	As req'd
Ultimate tensile stress	59,000 psi	57,000 psi	95,000 psi	90,000 psi	125,000 psi
Yield tensile stress	39,000 psi	42,000 psi	75,000 psi	70,000 psi	103,000 psi
Compression yield stress	36,000 psi	38,000 psi	75,000 psi	70,000 psi	113,000 psi
Ultimate shear stress	38,000 psi	30,000 psi	55,000 psi	55,000 psi	82,000 psi
Ultimate bearing stress (2 dia. edge dist.)	114,000 psi	108,000 psi	200,000 psi	190,000 psi	251,000 psi
Weight, lb/cu in.	.10	.10	.30	.30	.30
e = elongation	12–15%	12%	10–15%	over 17%	10%

Table 16-3 Material specifications for aluminum and steel alloys commonly used in aircraft construction

Aluminum alloys

Type	Bare sheet and plate	Clad sheet and plate	Rolled or drawn bars, rods, and wire	Drawn tubing	Extruded bars, rods, tubing, and shapes	Forgings	Castings
2014	H.T. to –T6 AMS 4029	QQ-A-250/3	QQ-A-225/4		QQ-A-200/2	MIL-A-22771 Type 2014	
2024	QQ-A-250/4	QQ-A-250/5	QQ-A-225/6	WW-T-700/3	QQ-A-200/3		
5052	QQ-A-250/8		QQ-A-225/7	WW-T-700/4			
6061	QQ-A-250/11		QQ-A-225/8	WW-T-700/6 Hydraulic tubing MIL-T-7081	QQ-A-200/8	MIL-A-22771 Type 6061	
6063					QQ-A-200/9		
7075	QQ-A-250/12	QQ-A-250/13	QQ-A-225/9	WW-T-700/7	QQ-A-200/11	MIL-A-22771 Type 7075	
356							QQ-A-601 Cond. T6
Tenzaloy							MIL-A-12033 Class 2

Steel alloys

Type	Sheet, strip, and plate	Condition	Maximum strength	Seamless tubing	Bars and forgings
1015-1025	Commercial	Mild steel	55,000 psi	MIL-T-5066	Commercial
4130	MIL-S-18729	Normalized Heat treat	<.187 thick–95,000 >.187 thick–90,000 150,000 psi max	MIL-T-6736	MIL-S-6758
4140		Heat treat	200,000 psi max	AMS 6381 AMS 6390	MIL-S-5626
4340	AMS 6359	Heat treat	260,000 psi max	AMS 6415	MIL-S-5000
321 & 347	MIL-S-6721	Stainless steel		MIL-T-8606	

by much cost, with welded tubular assemblies that have been brought up to this strength running the risk of looking more like pretzels than their original shape.

When higher strength is required for large and thick machined parts, 4140 or 4340 steels should be used. 4140 alloy can be heat-treated to 180,000 psi and even to 200,000 psi for certain applications, if such strength is absolutely necessary, while 4340 can be taken to 260,000 psi. Although steel parts heat-treated to these stress levels will exceed aluminum alloy strength-to-weight ratios, the 4130, 4140, and 4340 alloys are difficult to machine even in the annealed or normalized state and must undergo expensive and time-consuming heat treatment and inspection procedures before being bolted into the airframe.

So it is not surprising that welded tubular and sheet steel structures have been replaced by aluminum alloy for production aircraft. Although this approach also eliminates highly skilled welding labor, it has resulted primarily from the fact that steel structures are basically heavy assemblies that can be justified only for specialized applications, such as the pilot protection and serviceability required for Ag plane operation, as shown in Fig. 4-5.

Table 16-3 has been included to list various alloys and their uses as a procurement guide for readers interested in doing their own design or construction work. Whenever a specification is listed, the respective alloy is available in the form noted. When purchasing materials be sure to request documented certification of the alloy to specification because papers supplying this quality assurance information are required for FAA approval of any structural repairs or modifications. For example, when alclad 2024-T3 sheet is delivered, it should be accompanied by a notarized report stating the material was manufactured to the standards of specification QQ-A-250/5. But certifications will not normally be supplied unless requested.

For the record, alloy sheet is available in the following thicknesses (in inches):

- Aluminum: 0.016, 0.020, 0.025, 0.032, 0.040, 0.050, 0.063, 0.071, 0.080, 0.090, 0.125, and 0.190. Material of 0.250 thickness and above is available as plate.
- Steel: 0.025, 0.032, 0.040, 0.050, 0.063, 0.072, 0.080, 0.090, 0.100, 0.109, 0.125, 0.160, and 0.187. Heavier material is available as plate, while 0.025-in 4130 sheet is becoming very difficult to obtain.

Local metal supply sources should be contacted for additional information concerning aluminum and steel alloys. Free data is usually available, a typical example being the useful Alcoa handbook.[2]

Fiberglass, foam, and similar materials were briefly covered in chapter 12 under the heading of composite structures. These synthetic materials have particular appeal for simplified homebuilt aircraft, having reached maximum application in the virtually 100% synthetic Rand KR-1 and KR-2 designs, as well as Burt Rutan's VariEze of Fig. 3-5.

Sailplane designers are increasingly turning to fiberglass and foam construction, particularly for higher-performance models where a smooth exterior finish is of primary importance to drag reduction. Due to the absence of engine vibration, fiberglass structures stand up very well for this type of aircraft; although in off-field landing accidents glass offers much less protection than sheet aluminum.

If existing synthetic materials and construction techniques are used, powered aircraft designed for FAA approval will experience a structural weight penalty accompanied by considerable reduction in useful load. Hopefully, new composite materials and methods can be introduced to provide the same level of structural integrity and reproduction offered by sheet metal—and at similar cost.

In the meantime, it is possible to design an aluminum or steel alloy tubular fuselage structure capable of carrying primary loads and then enclose this assembly within a contoured envelope of fiberglass covered polyvinyl chloride foam. This method of construction permits integral fairing of wing fillets, canopy supports, tail surfaces, etc., as well as buried antennas, all within a polished, smooth skin that can be readily removed for access and inspection.

Unfortunately, faired tubular construction is not suitable for working seaplanes because of water loads, floating debris, thin ice (which can act like flying slate to penetrate the covering), and the rigors of docking and loading. But it offers drag and assembly advantages for the next generation of high-performance, efficient landplanes, some of which are now being designed for metal and foam construction.

Protective metal coatings

Typical of the compromises arising in most engineering work, the advantages offered by a metal structure are somewhat reduced by the threat of corrosion. Although aluminum alloy does not rust, it can surely corrode, particularly in the presence of a salt-laden atmosphere. And frequently aluminum corrosion is not evident until well advanced—similar to termites working in floor beams and sill plates. Steel is a bit more honest; corrosion is immediately evident as rust streaks leach out over everything nearby. Because neither form of corrosion can be tolerated for primary airframe components, methods have been developed to protect metal surfaces both before and after assembly.

Aluminum alloy sheets can be made with a clad coating on both faces and receive no further interior surface finish during the life of the airplane. Or more protection can be added in the form of a chromate conversion coating, a type of chromic acid chemical dip treatment generally referred to by the original trade name of "alodine" or "alodining." Bare aluminum should be alodined prior to painting, not only to protect the material but also to provide a better paint bond, which, in turn, further protects the material.

All aluminum alloy parts for seaplanes should be fabricated from clad sheet and then alodined before painting to provide maximum protection from saltwater corrosion. Alodine touch-up kits are available for coating local scratches, abrasions, or surface repairs and should be part of the maintenance inventory for every metal airplane.

Steel fittings and small sheet parts should be cadmium-plated to protect the metal surface from environmental corrosion. Cadmium plating is an electrochemical process during which the exterior surface of steel parts is coated with a cadmium deposit varying from two ten-thousandths to five ten-thousandths (0.0002 to 0.0005) of an in. in thickness. A 0.0003-in coating is standard in the lightplane industry and is referred to as Class 2, Type I Cadmium Plating to specification QQ-P-416. The cadmium surface formed is relatively soft and tends to corrode before steel will, thereby covering surface scratches and preventing steel from rusting.

Recently more environmentally acceptable finishes than cadmium plating have been developed for steel surfaces. Among these are thermal spray and dry coating processes, but a primed finish over cad plate still remains one of the best corrosion protections for steel.

Complex steel parts such as weld clusters and large assemblies should not be cad-plated. Chemicals used in plating can become trapped in recesses or weld laps and then slowly but surely eat away on the structure, accomplishing just the reverse of what was expected. To protect the surface, steel assemblies of this type should be sandblasted or wire-brushed and then immediately painted with primer.

When parts to be cad-plated have 180,000 psi or greater strength, it is necessary to have them stress relief heat-treated prior to plating. This step is required to prevent surface embrittlement and possible development of fatigue cracks. Since small aircraft steel parts are rarely raised to such high stress levels, plating embrittlement usually is no problem. However, spring steel wire is of much greater strength and to prevent early failure must be properly stress-relieved when cad-plated, usually by a baking procedure.

Steel tubing weld assemblies require some form of internal protection to prevent moist air trapped during welding from steadily corroding the structure from the inner to the outer surface. Unfortunately, many weld assemblies are not properly protected and subsequent repair becomes extremely costly, to say nothing of the potential danger of an in-flight structural failure occurring before corrosion is noted.

Although many commercial solvents are offered for steel protection, the best method for welded tubular assemblies seems to be a raw linseed oil internal rinse. To be effective, cluster joints must contain holes permitting internal flow of oil through the assembly, or be drilled in nonstructural areas for external draining. Once the assembly has been flushed of oil, any external drain holes must be sealed with plated steel drive screws (such as Air Force-Navy Standard

Hardware listed in AN535). Boiled linseed oil or pure linseed oil is not necessary; a commercial grade of raw oil is equally effective and much cheaper. Some users recommend heating the oil to about 180°F to drive out moisture and to permit the applier to follow the oil flow through the structure by touching the exterior surface and feeling heated areas.

Surface finishes

Whether of aluminum or steel, surfaces exposed to the weather should be coated with primer paint after alodining or cad plating. Primer is a fast-drying finish that can be applied by brush, spray, or dip, leaving a zinc chromate or epoxy base for finish painting. Various types are available, with zinc chromate being somewhat lighter and cheaper than epoxy and, therefore, the type most commonly used for priming small aircraft.

If an untreated aluminum alloy surface is to be painted, a "wash primer" coating is recommended prior to spraying any final finish. Wash primer is superior to zinc chromate primer for this application because it tends to etch into bare and clad metal surfaces, providing an excellent bond for finish painting. Of course, all this doesn't come easily because wash primers are two-part compounds that tend to "string" or congeal if not properly mixed. But the added protection obtained from wash priming is worth the effort, although guidance from your local paint supply source is strongly recommended prior to application.

Scratches or gouges in a fiberglass structure should always be given immediate attention, because it is possible for water to work between the glass layers. This moisture can either expand as trapped steam when the airplane is parked in intense sunlight, or freeze and crack the structure locally during exposure to low ambient temperatures. Fiberglass repair kits are available for quick touch-up jobs and should be used as soon as possible after surface defects are noticed.

Automobile body filler will smooth metal and fiberglass surfaces when local dents or gouges cannot be worked out, and should be applied in the same manner as minor auto body repairs are made at home. A better and lighter material is a form of fiberglass filler known as Feather Fill, available from the source noted at the end of this chapter.[3] This filler can be sprayed or troweled in place on metal or fiberglass, sanded smooth, and then finish painted. The smoother the sanding job the better the surface; unfortunately, much elbow grease is essential to all good finish work.

Finish paints come in about as many brands, trade names, and different base compounds as there are colors. Although enamels and lacquers have been around for years, the enamels have recently branched into acrylic and polyurethane bases as well as the old standard type, each with its own group of enthusiasts. Polyurethane might be the easier to apply because of its fairly long pot life, but both acrylic and polyurethane enamels are two-part paints requiring an activator plus the use of stir pots during spraying. Either can be hot-sprayed for a fast drying finish and will provide excellent protection with a high gloss. So take your choice.

For fast backyard or hangar repairs, lacquer is easy to apply and offers adequate protection when the base coat is sound. Plain old 4-hr drying enamel will probably last longer, offer a higher gloss finish, and can be applied over any paint. Remember that lacquer cannot be applied over enamel; it will attack the surface and immediately develop a crinkle finish.

Although the choice of paint will probably depend on the color match available, finish paint provides important protection for the airframe. Any scratches or local peeling should be covered as soon as possible, even if primer is the only finish readily available.

Seaplane protection

Seaplanes likely to operate in saltwater must have special protection in addition to exterior paint. When possible all aluminum alloy parts should be fabricated from alclad sheet, alodined, and primed. Control cables should be of 7 × 19 stainless steel, specification MIL-C-5424, for maximum life and strength. All exterior cables should be coated with Par-Al-Ketone, and touched up as often as necessary to maintain a continuous coating.

All interior and exterior bolt heads, nuts, and bolt ends should also have a brush coating of Par-Al-Ketone, with interior bolts and nuts given an additional seal of Vaseline if located down low in bilge areas. This represents a lot of work, but so does replacement of badly corroded fittings. And stainless bolts or screws, contrary to popular belief, are not an acceptable substitute for cad-plated steel AN (Air Force-Navy Standard Hardware) bolts in aluminum alloy structures exposed to saltwater. I learned this the hard way a few years ago when I finally had to chisel and drill out a steel bolt that, in one sailing season, had become frozen (corroded) into the extruded aluminum mast of my ocean racing sailboat. Galvanized dip-plated steel parts behave best around aluminum and saltwater, but the plating is so soft and irregular in texture that galvanized bolts are not acceptable for joining highly stressed structural members.

With constant attention and a quick check of crucial areas after each water flight, the special problems of saltwater operation should not demand excessive maintenance. Probably the most effective corrosion inhibitor of all is a thorough wash down with fresh water immediately after each briny dip. If washing is not possible, fly to a nearby freshwater lake or river and run through two or three landings (gear-up, if an amphibian); this is not as satisfactory as a good hosing, but much better than nothing.

When all is said and done, keeping your airplane clean and polished contributes the most toward a lasting finish; and of course, the smoother the surface the lower the skin friction drag, and the better the performance. A polished airplane also reflects pride of ownership.

So you gain all around if you can treat routine and finish maintenance the same way you would any pleasant hobby. Then the only way you can improve caring for your airplane is by keeping it hangared. Being under cover is the ultimate way to increase the life of the entire airplane, in addition to assuring a lasting finish.

Endnotes

[1] AC 43.13-1A, pp. 132-139.

[2] *Alcoa Structural Handbook*, Aluminum Company of America, Pittsburgh, PA.

[3] Feather Fill is available in pint, quart, and gallon quantities from: Fibre Glass-Evercoat Co., Inc., 6600 Cornell Road, Cincinnati, Ohio 45242.

17

What is certification?

I'm sure we've all seen publicity releases concerning a proposed new airplane design, including a responsible first flight date plus the notation that FAA approval is expected about 6 months thereafter. And quite likely the person or management group responsible for each announcement was completely sincere and expected certification by the date indicated.

The unfortunate part of all this being that aviation people tend to be overly optimistic, and they frequently have their best dreams shattered when aircraft development programs drag on and money drains away. This partially explains why so few new aircraft that do fly ever obtain production approval, with less than 10% of such projects ever receiving a TC, the Federal Aviation Administration Design Type Certificate.

After having lived through six type certifications with all their attendant delays, frustrations, and financial demands, I thought it might be helpful to pilots and budding entrepreneurs alike if what goes on during the certification process was discussed, including why so much time and money fly by between the first design sketches and final approval by the FAA.

Although most pilots are aware of the FAA requirements covering private aircraft design under Part 23 of the Federal Air Regulations (known as FAR Part 23), very few people fully realize the detailed work and procedures necessary to obtain a TC. And after reading this description of the process you might well wonder why any aeronautical engineer would go down the certification path a second time, to say nothing of five or six times.

The market arena

The general aviation industry is unique in that it tends to sell almost entirely to itself. That is, aircraft are designed, built, and sold to people interested in flying; powerplants and electronic equipment must be specifically developed for aircraft use; personal aircraft airports and fixed-base operations develop and expand only in communities having at least one enthusiastic and organized aviation group; and all this work originates with people personally interested in and familiar with the benefits of flying. As a result, the various segments of general aviation are closely interrelated; when one section experiences a reduction in sales volume, the entire industry soon follows, fanning the fire.

Since private aircraft and related flight training are luxury items for the majority of owners and users, our industry rapidly responds to negative factors in the economy. When national income falls or major business profits decline, the resulting belt-tightening process rapidly squeezes out private flying. All of this means designers of small aircraft must be both attentive and responsive to current and developing industry trends, for these are potential customer requirements that will influence the acceptance of any new product.

Development of design requirements

A new aircraft design usually originates through the following procedures, either singly or in combination:

- An extension or improvement of an existing design, such as increased wing area and useful load, or increased horsepower, cruising speed, and useful load.
- In response to dealer requests arising from customer inquiry and/or operating experience. Hopefully, backed by dealer orders for the new model.
- Through designer concept of a new type of configuration for which it is believed a market exists. This approach can be dangerously costly unless backed by firm orders from others willing to share the development risk and investment required.
- A special configuration designed to order. This can be an experimental designer's ideal situation, as he or she will be paid for developing a new airplane. Because production runs are unlikely for special models, activity of this type is usually restricted to individual designers or very small aircraft companies.

Major details of any new airplane are first roughly sketched to support equally rough weight, balance, and performance estimates. Once the resulting concept receives customer or management approval, preliminary design work is started in exacting detail.

The concept phase

From the basic preliminary sketches developed to launch the new project, the design engineer next evolves the final airplane configuration, meanwhile balancing and juggling allowable weight items, budgeted development and production costs, tooling vs. labor-hour production trade-offs, and FAR Part 23 detail design requirements.

One of the first layouts will be an inboard profile, or sideview, showing the location of all systems and major structural members. This drawing requires considerable effort because it must realistically position passengers, equipment size and location, control system components and movement, instrument and radio location and space requirements, landing gear position and correct size of supporting structural members, tail clearance for maximum lift when landing, major structural bulkheads and longeron sizes, powerplant installation, pilot and passenger vision angles, etc. Such basic requirements as FAA ground clearance for the propeller as well as safety belt and shoulder harness attachments must also be determined. Each of these items requires detailed design study and usually some compromise discussion before the design is frozen in final form.

All components must be checked for proper weight and center-of-gravity location at normal load, and also for maximum forward and aft center-of-gravity travel with disposable items in place and out to make sure the balance range falls within acceptable aerodynamic and ground handling limits. Of course, while all this is going on, company personnel feel obliged to comment on each line set down, even though such work might be very preliminary. Fortunately, many favorable ideas result from random comments, so all suggestions should receive studied attention—which time has shown to be the safer and more socially acceptable, though difficult, approach.

When the weight, balance range, structural arrangement, and aircraft outline appear acceptable, the design is considered fairly well set, subject to at least an informal mock-up of the cabin arrangement. For some reason, location of the seats, control wheel or stick, and rudder pedals never seems to transfer properly from paper to three dimensions, so some rearrangement of these items is usually required.

Once the configuration is approved, the decision to produce detailed design drawings follows. At this stage, an aircraft company becomes committed to expenditure of a sizable sum if the design is to be taken through FAA certification; a decision frequently made, incidentally, without any guarantee of sufficient production to realize at least the return of development capital.

For this reason, normal business financing sources are understandably reluctant to provide working capital unless the company has more than adequate collateral to cover any loans required. This means that if a small company does not have adequate net worth or a large cash flow developed by other product lines, it must find sources of risk capital before undertaking a new design. And if the new airplane is not successful, all responsibility falls on the designer's shoulders.

Assuming the necessary financing is eventually obtained, you can then proceed along the TC approval road by advising the FAA of the new program. Parallel with detailed layout and design drawings, the engineering department must develop the applied loads that actually establish structural requirements. Part 23 specifications set forth these loading conditions in great detail, and for some items, such as tail surfaces, it might be necessary to calculate three or four design conditions before the crucial set of loads is determined. Depending on the size of the airplane, and whether or not it is subject to water loads and/or twin-engine operation, as many as 100 pages of applied loads data might be necessary. Of course, these end calculations might be backed by another 200 pages of detailed analysis; so considerable engineering effort is required to prepare this basic information.

The applied loads are subject to close review and must be approved by the FAA because the safety of the resulting design depends on the thoroughness and accuracy of these calculations. As a result, considerable time can pass before the applied loads are accepted. During this period the prototype (first experimental) airplane and tooling are growing at a fast pace, and any change in load requirements can be both time-consuming and costly. Again, the designer feels the impact of any modifications required.

Design and construction of the prototype

As detailed tooling for the new design is completed, one or two sets of parts are run off and prototype assembly begins. If most of the structure is to be substantiated by detailed mathematical analysis with a minimum of static test work, usually only one set of parts need be made; if extensive static test work is required by engineering, additional parts and assemblies must be fabricated.

Some designers prefer detail structural analysis to static tests of the basic components because the margins of safety obtained through such analyses are available for future airframe growth and development. Also, complex static test programs can require so many calculations and test fixtures that this route proves more costly than detailed analysis. However, the FAA does require certain mandatory static tests such as

- Wing panel assembly
- Wing torsional stiffness
- Flutter investigation
- Fuselage assembly
- Control system
- Control surfaces
- Tail surfaces
- Landing gear drop test
- Engine mount
- Fuel tank endurance tests (and oil tank, if used)
- Pilot and passenger seats and belt attachments
- Baggage compartment floor

Static tests to ultimate load requires structural assemblies built for that purpose because any items tested to ultimate load (that includes a factor of safety of 1.50 times the design load value as discussed in chapter 2) cannot be flown and must be scrapped. Further, any future increases in gross weight or loading of any component proven by static test alone will require retest. So the load analysis better be correct when tests are run or they will have to be rerun, incurring a further loss of time and money.

As a matter of policy, all drawings, design analyses, and static test work are subject to detailed review by FAA regional engineering personnel. For some items the FAA will accept design approval submitted by DER consultants. The DER is a Designated Engineering Representative appointed by the FAA as competent to perform engineering work on its behalf. A DER can be approved for airframe stress analysis, flutter analysis, systems and equipment design, powerplant design, flight test, or flight data analysis, and is of great assistance in preparing and expediting engineering work for TC design acceptance. However, even with DER approval in hand, FAA engineering reserves the right to review and disapprove any and all items, although such occurrences are extremely rare.

During the fabrication of all airframe parts and prototype assembly, company inspection procedures and records must be established and maintained in detail to substantiate the quality of work completed. FAA inspection will be provided by the assigned MIDO (Manufacturing Inspection District Office) or similar representative who should be brought into the project at the earliest possible time. It is most difficult to inspect components of a fully assembled wing, fuselage, or tail section; and even more difficult to dismantle assemblies to permit FAA inspection. Without FAA inspection approval the prototype cannot be flown by FAA test pilots, and therefore cannot be certified.

When the prototype is completely assembled, certain static tests must be performed and FAA approved prior to initial flight test; these include the control system and surfaces, wing torsion, and flutter investigations.

Following these demonstrations, and providing they are supported by drawings and parts manufactured to FAA inspection approval, the new airplane will be given an experimental airworthiness license permitting flight by company pilots. It should be noted that all detailed drawings and structural analyses need not and probably will not have received FAA approval at this time because flight work normally indicates a need for some minor or major changes.

At this point the FAA regional office might elect to convene a Preliminary Type Board for review of the new design. The purpose of this meeting, held at the manufacturing plant, is to submit the prototype to close study by all regional department heads. Following this review the board will prepare a list of recommended detail modifications considered necessary to meet FAA specifications. This is an important step in the TC procedure and indicates the program is moving along very well, regardless of the length of the resulting modification list.

The flight test

The experimental prototype will be expected to remain in the vicinity of the home airport for the first 10 hrs. of flight. Test investigations must be conducted over fairly unpopulated areas (presumably to prevent dropping out of the sky onto housing developments), and no persons can be carried except those required to obtain flight data.

During company conducted flight tests, information will be gathered to substantiate compliance with Part 23 detailed flight requirements. The FAA has a flight test form most useful in completing this work that covers stability throughout the flight envelope, design dives beyond the red line limit to be used for production aircraft, fuel system operation, engine cooling, ground (and water) handling, climb rate, and spins—to name only the major areas.[1] Different interpretations of FAA flight requirements as well as variations in pilot opinion regarding handling characteristics can make the flight test period extremely frustrating at times. However, once company pilots agree that the airplane configuration meets all FAA flight requirements as understood, the design is considered frozen for TC approval.

Following that decision, the entire design data package must be submitted and approved before an FAA test pilot will be permitted to evaluate the airplane. Upon completion of regional engineering design approval, the FAA issues a Type Inspection Authorization (TIA) requiring their inspector to review the prototype for conformity with the approved drawings. This covers detail compliance with all drawings and manufacturing procedures, including any flight required modifications such as area changes, fillets, dorsal fins, revised cowling, equipment changes, etc. When this inspection is satisfactorily completed and documented, FAA flight test evaluation can begin. This level of achievement is a major step toward certification and one realized by a very small percentage of new design starts.

The FAA regional test pilot staff will assign three or more pilots to evaluate company flight data, with one person designated as project pilot. Assuming company evaluation has been conscientiously conducted, this final phase of certification should run smoothly. Unfortunately, differences in interpretation of test requirements and/or results can arise, frequently prolonging flight approval. If structural or equipment changes result from FAA flight evaluation, another TIA must be issued and another FAA inspection completed before FAA flight work can be resumed. So this phase of evaluation can run into a series of 2- to 4-month delays for a small and seemingly uncomplicated design, as well as becoming very expensive with flight test work costing close to $200/hr. and salaries to be paid as time goes by.

In addition, if an initial run of production aircraft is coming down the line, any delays in final TC approval can tie up large sums of scarce capital, while all flight required changes must be retrofitted to aircraft in process. For obvious reasons, a small company is well advised to start production after receiving FAA Type Certificate approval, and not before.

Phases of FAA approval

Once the design TC has been granted, the company holding the TC has 6 months in which to establish effective manufacturing and inspection procedures for the new model, as demonstrated by essentially identical empty weights and flight characteristics between successive production aircraft. FAA inspection personnel will assist in defining crucial items and required procedures. At the end of this period, a new aircraft company will be surveyed by the FAA and any deficiencies noted. Inspection and manufacturing process procedures must then be documented by the company in accordance with a mutually acceptable format. Following approval of the resulting procedures manual, a new company is granted an APIS (Approved Production Inspection System) certificate, while an established airframe manufacturer has the existing Production Certificate (PC) amended to include the new airplane. And this is all part of the design TC procedure because without an APIS or PC in process or approved, it is not possible to sell production units of the newly TC'd design. After an APIS is granted, the manufacturer must continue to refine fabrication and inspection procedures leading toward PC status as soon as possible.

In summary, the five crucial steps for obtaining production approval of a new airplane design are

1 Notifying the FAA regional office that a new design program will be initiated. The FAA office then assigns a program code to the proposed design. General arrangement drawings, as well as preliminary weight and balance plus performance data should be submitted at the time FAA consideration of the new design is requested.

2 Preliminary Type Board review.

3 Type Inspection Authorization, which represents completion of the engineering data review and approval of the aircraft design for conformity inspection prior to FAA flight evaluation.

4 The award of the design TC.

5 Development of manufacturing and inspection procedures covering the new design to assure consistent quality of production aircraft, resulting in the award of APIS or PC certificates.

These five steps normally require 24 to 36 months of concentrated effort, accompanied by considerable engineering ability and much capital. So you can see why new aircraft are safe, but expensive.

And regardless of whether a new airplane is being approved under FAR Part 23, the new Small Airplane Certification Compliance Program, Primary Category, or the Joint Airworthiness Requirements (JARs), the procedural steps are similar, only the extent of detail varies.

In this regard, interest has been developing over the past ten years in the possibility of simplifying Part 23 design regulations for truly small aircraft intended only for personal business and recreational flying.

While such aircraft might not be offered for lease or hire, any regulation change was requested to include the right of the builder/owner to perform required service and maintenance after receiving instruction from the factory producing the kit or production airplane.

In addition, it was petitioned that the revised regulation permit flight instruction in two-seat or larger versions of such aircraft.

We now have the necessary FAA approval for such simplified design specifications; four categories as listed in Table 17-1, although it should be noted that ultralight aircraft are not subject to and cannot obtain FAA Type Design Certification approval.

The Small Airplane and Primary categories appear to be the best defined and most popular of these new regulations, with a number of small aircraft being certified for production at this time.

The Primary Category Final Rule carries some rather interesting and somewhat vague design regulations.

To quote from this rule as described in FAA Document Number 23345, as amended and dated September 9, 1992, in addition to the key criteria listed in Table 17-1;

(1). The final rule permits kit aircraft supplied by the holder of a production certificate to be assembled by another person under the supervision and quality control of the production certificate holder.

Alternately, if a kit aircraft supplied by the holder of a production certificate is assembled by another person who is not under the supervision and quality control of the production certificate holder, the completed aircraft is eligible only for an experimental airworthiness certificate.

(2). The FAA agrees that the development of certification standards by the private sector represents the most productive and cost-effective manner of streamlining the certification process.

The development of airworthiness design criteria by the private sector would be similar to the FAA's Technical Standards Order (TSO) authorization program.

The simplified type certification process envisioned for primary category aircraft is expected to draw heavily from airworthiness standards already in the regulations, existing delegation procedures, and statements of compliance made by applicants for type certification.

(3). Section 21.184 (c) provides for an aircraft with a standard airworthiness certificate to obtain a primary category airworthiness certificate.

(4). Primary category aircraft may be used for flight instruction under certain restrictions.

Above paragraph (1) advises that any homebuilder who purchases a kit for assembly of an FAA approved primary category airplane and proceeds to complete fabrication without factory supervision, such as special training or assembly at the factory under direct supervision, will become the owner of just another experimental licensed homebuilt, not a primary category airplane.

The second quotation clearly indicates that the path to primary category type certification might not be well defined, but instead subject to local FAA Aircraft Certification Office preferences.

Table 17-1 Simplified FAA Regulations

TC category	FAR specifications	Maximum gross weight	Maximum number of seats (including pilot)	Maximum landing speed
Small airplane certification compliance program	Simplified Part 23 (similar to primary) fixed gear only	3000 pounds	4	61 knots 70 mph
Primary (Docket no. 23345)	Final rule–09/09/92 Part 21.24, 21.31, 21.163, 21.184, 21L Part 43, App. A (c) (30) Part 91.325	2700 pounds	4	61 knots 70 mph
JAR–VLA Joint airworthiness req'ts–very light aircraft	FAR 21.17	1654 pounds	2	45 knots 52 mph
Ultralight aircraft	Part 103 (No TC available)	*254 pounds *350 pounds	1 2	24 knots 29 mph

*Empty weight

Notes:

1. Certified primary category aircraft may be offered in kit form as well as completed production aircraft.

2. Aircraft may not be pressurized.

3. These regulations require a single, normally aspirated engine located on the airplane centerline.

4. Engines and propellers may be approved for all above categories via the design specification, including Parts 33 and 35.

5. Carrying passengers for hire is prohibited, although flight training is permitted for primary category aircraft under certain conditions.

However, while the design specification can take some time to develop and receive approval of all detailed items, it is expected that the final regulation for a particular primary category airplane configuration will be somewhat less complex and costly than current FAA Part 23 requirements.

The significance of paragraph (3) lies in the somewhat unusual condition that a production airplane certified under Part 4b, CAR3, or Part 23 can have its airworthiness certificate exchanged via STC for a primary category airplane airworthiness certificate provided the design meets the weight, seating, and landing speed criteria of Table 17-1.

Why would an owner wish to do this? To obtain the right to perform much of the required service and maintenance work, provided the proper training has been received and the airplane will not be used for hire.

While conditions for design and acceptance approval are not clearly defined and have not been extensively tested at this time, it is expected that approval and operation under the Small Airplane Certification Compliance Program and JAR-VLA requirements will be similar to regulations published for primary category aircraft.

One of the earliest applicants for Certification Compliance approval is the interesting, high-performance Lancair ES. A four-place, 2900-pound gross weight design, the ES is intended for series production following FAA Type Design Certification (see Fig. 17-1).

Some closing thoughts

Although I do not agree with all Part 23 requirements, and particularly object to preparing fail-safe design and fatigue analyses for small aircraft used only a few hundred hours a year, there can be no argument that current FAA flight specifications produce a very stable, easily handled airplane. As flight proficiency requirements were lowered over the years to accommodate students and occasional pilots holding only a basic private license, aircraft design standards were raised to provide forgiving wings. And this is a proper philosophy—flying should not be a constant test of pilot skill if engineering can make it otherwise. The airplane should be developed to serve the average pilot, not the pilot to serve the airplane by virtue of superior skill.

On a constructive note, every designer has the opportunity to study and incorporate ways to eliminate pilot error items before the airplane flies. Based on more than 50 years of design experience, I believe many so-called pilot error incidents are more the fault of the designer than of the pilot. In fact every aircraft designer has an implied responsibility to eliminate pilot error at the source by "designing out" opportunities for error. It is my sincere hope this book will make a positive contribution toward achieving that goal, while also adding to every pilot's understanding of how his or her airplane is designed and why it flies as it does.

Photo by James Lawrence for Lancair International

17-1 *The new four-place Lancair ES is a candidate for FAA approval and production under the new certification compliance program.*

Endnotes

[1](a) "Flight Test Report Guide;" FAA Form 8110-11 prepared by the FAA.

(b) "Flight Test Guide for Certification of Part 23 Airplanes; AC23-8A," dated 2/9/89.

Available from the Superintendent of Documents; order SN050-007-00817-1. Price $12 as of 12/93.

Index

Illustration page numbers are in **boldface**.